Nuclear Structure from Sphere Stacking

Written during June, 2021

Published by Alan Folmsbee on KDP
Self-published using Amazon Kindle Direct Publishing
The paperback edition was copied to the Kindle eBook version

Copyright © 2023 Alan Charles Folmsbee
All rights reserved

Bristol, Connecticut, USA

First edition, June 14, 2021
Second edition September 8, 2023

68 figures in 108 pages with 23,579 words

Contents

Preface 5
Introduction 7
Rules of a Static Nucleus Theory 9
Chronological order of the discovery events 15
Highlights of results 20
Matter and vacuum 23
Iron versus copper 27
The Periodic Table of the Elements 41
References Section A 46
Carbon is unique 47
Chromium from Argon 54
References Section B 73
Promethium and technetium 74
Platinum and vanadium catalysts 80
Evidence of the correctness of this theory 85
Electrons forcing protons into line 89
Abundances of the elements 94
Origins of the cubic lattice 98
Conclusion 101
References Section C 103
Index 106
About the author 108

Preface

The subject of nuclear structure has been discussed for years, but little progress was made. Most scientists expected the nuclei of the 118 elements to have complicated explanations that needed advanced algebra. Dynamically changing shapes were predicted. Iron was expected to be shaped like a sphere. Now, it has been discovered that the shape of the nucleus is much simpler than expected. This book gives full details about the series of choices that resulted in a simple geometric theory for the shape of the nucleus. Reasonable theories, like electrostatic repulsion were used. The element iron was chosen to begin the evaluations of possible shapes. The readers of this book can be high school graduates who got a grade of A in physics. Also, the readers are expected to be college students and graduates. You readers will be treated to a new Periodic Table of the Elements (Fig. 23) that can be used to guide future collider experiments. Theory is ahead of experiments for a change. This is a new situation for investigations of nuclear structure. Baryons are protons and neutrons.

Geometry is the appropriate branch of mathematics for progress in the understanding of nuclear structure. Three dimensional sculptures were required to clearly envision the shapes of the elements. This book only has two dimensional pages, so it will be difficult to cause much certainty for the reader to enjoy. Models can be made by the reader using **The Rules** that the author used. The reader can also obtain sculptures of the elements that are made by other people. Three dimensional printers are a good source of the models and a website called thingiverse.com already has the data files that were provided by the author.

The references are in three groups: A, B, and C. For example, [10C] means the tenth reference in group C, near the back of the book. Group B is after the chromium paper and Group A references cover the first chapters of this book.

The second edition has many corrections and the Figures are now numbered 1 through 68. New images are used to replace older images that were too dark. The Periodic Table of Nuclear Structure is now only shown one time. It covers two pages so it has bigger silhouettes of nuclei. The table of contents now has the page numbers included. The Rules for building models of the nuclei of all chemical elements are listed only once in this edition.

Introduction

The shapes of crystals are known to reflect the shapes of atoms, and their electron orbitals. The probabilistic formulation of the Schrödinger equation provides a cloudy picture of the shapes of atoms and molecules. Chemists have used those blurred guidelines without knowing the shapes of the nuclei. Now that this book is revealing the nuclear structures of all elements and The Rules that determine them, chemists will begin to achieve a higher level of precision in their simulations than were ever possible before. The "charge distribution" has been important for chemists and physicists for many years. They map the electron position estimates, but not the proton charge distribution estimates, because there were none. Today, that has changed. All elements' proton positions and neutron positions can be known exactly from the new theory.

Catalysts are important to modern industry. Platinum is a good example of a catalyst. If methane is blown across a fine wire of Pt at room temperature, the gas will burn. Vast wealth was spent to use that catalyst, but nobody knew why platinum was so good. Attempted explanations used concepts like energy conservation and the square root of minus one to explain platinum. Those explanations were unsatisfactory. This book will describe the meaning of charge and why the locations of protons in platinum make a hook, a ring and an axis for charge distribution. Vanadium is also shown, following the new Rules, to also have a ring of protons, a hooked line of protons inside the ring, and a single proton on the axis of the hook and ring.

Ferromagnetism was, in the past, attributed to electron spins, with little mention of protons. Now, the real reason for ferromagnetism is provided. Iron has two rings of protons that are coaxial with each other. On the axis are two isolated protons. Each ring has 12 protons. Gadolinium has 18 protons on each ring. Chromium has 10 protons on each of two non-coaxial rings. Neodymium has two rings of protons, but it lacks the isolated proton on its axis, so it is nonferrous. Nd provides flux vortexes that can be gathered by iron's flux vortexes, so the B field is quadrupled in the famous neodymium magnets.

Carbon is different from all other elements. It is the element of life, of diamonds, and of graphite. In the past, science did not say why carbon is so excellent, except for praising energy conservation and the square root of minus one. But this book shows the solution to that mystery. Carbon is the only element that has protons that define two parallel planes that are symmetrical. This introduction should increase your curiosity about the static nucleus theory that defines the pyramidal cube nuclear structure.

There are 19 foundation elements, upon which, most other elements are built. Those other elements are called incremental elements if they are based on a foundation element. **A foundation element is defined** as having a simple cubic stacking of protons and neutrons (baryons) in the core and having a pyramid of baryons piled onto each face of the cube. That shape allows survival, of recently fused elements, for a time approaching eternity.

Rules of a Static Nucleus Theory

The rules that determined the nuclear structure of iron were extended to become 19 Rules that were applied to all elements. This summary does not explain the rules, here. See the justifications for the Rules in [22C], where 6 pages explain the geometric reasoning and considerations of the properties of the chemical elements that were used. A foundation element has a nucleus with a cube of baryons and six pyramids of spherical baryons that cover the six faces of the cube. An incremental element adds baryons on the foundation nucleus. Protons and neutrons also called baryons. Z is the atomic number and A is the mass number of a chemical element's nucleus.

The 19 Rules contain a list of 19 foundation elements.

The Nineteen Rules of Nuclear Structure Using the Pyramidal Cube Theory

Rule 1: There is a simple cubic lattice of protons and neutrons at the core of each element that has a Z that is greater than five.

Rule 2: Protons in the cube are far from each other as if electrostatic repulsion is in effect.

Rule 3: The six faces of the cube are armored by pyramids of protons and neutrons.

Rule 4: Protons outside of the cube tend to form lines of protons as if electrostatic repulsion is not true in all three dimensions.

Rule 5: There are 19 foundation elements upon which the 90 incremental elements are built. The 19 foundation elements are:

carbon, oxygen, neon, phosphorus, argon, iron, germanium, krypton, zirconium, cadmium, xenon, cerium, hafnium, tungsten, polonium, radon, uranium, mendelevium, and nihonium.

Rule 6: The shapes of foundation elements do not depend on protons being different from neutrons. Both are treated equally, as baryons, to define the silhouettes and 3D shapes of each element.

Rule 7: Four sides of the cube have pyramids of the same shape (axial symmetry), for foundation elements. The top and bottom pyramids can have different sizes. All of the side pyramids are equal in size and shape. Rotations of pyramids do not need to be identical when positioned on the four side faces of a cube.

Rule 8: Pyramids should be rotated to avoid creating a three-way intersection of lines of protons. Some elements cannot avoid that structure, like promethium and nitrogen.

Rule 9: Incremental elements have added nucleons on the exteriors of foundation elements to fill the gaps between pyramids. There are 90 incremental elements based on the foundation elements. Nine elements are not based on a foundation element. They are H, He, Li, Be, B, Tc, Pm, Pa, and Og.

Rule 10: An incremental element is assembled by first placing the neutrons into the deepest pits of a foundation element and then adding one proton into the deepest pit where protons tend to form lines of protons. If a line cannot be formed, the added proton can go anywhere that does not join 3 protons together in a triangle. If that is not possible, a proton can go anywhere.

Rule 11: Light elements have a sparse allocation of protons near the center and a denser allocation of protons near the tips of pyramids.

Rule 12: Pyramids can be up to six layers thick.

Rule 13: Contraction of pyramid bases occurs increasingly with heavier elements. A six-layer pyramid can rest on a five-layer contracted base, which can reside on a four-layer contracted base, which can reside in a three-layer cube, nestled into a stable arrangement.

Rule 14: Sphere stacking for a pyramid does not need to nestle into pits of a cube and the pyramid can be stacked onto a cube vertically. For example, in oxygen, a two-layer pyramid can be stacked onto a two-layer cube.

Rule 15: Pyramids can have lines of protons plus additional protons at the corners of pyramids to achieve the Z atomic number that is known by standard science.

Rule 16: Symmetrical arrangements of protons are preferred over non-symmetrical structures. The same is true for neutrons. The two-layer pyramid sets the example in iron. The cube-2 and cube-3 are also symmetrical in their allocations of protons and neutrons.

Rule 17: Each nucleus is shaped to provide the isotopes with A and Z which were established from old experiments for established physical tables.

Rule 18: Each proton has one electron paired with it using a line of flux. This drives multiple protons into a single line of protons that touch each other.

Rule 19: The longest distance from each neutron to a proton is one diameter of a neutron.

Justifications for the Rules are given in the large reference book [22C] <u>Charge distributions on the nuclei</u>.

Figure 1: iron-57 using white protons and dark neutrons

Notes on the rules of nuclear structure

The stability of heavy elements is due to armor on the faces of the cube. This structure is called the Face-Armored Cubic lattice. This is a terminated crystal lattice (Fig. 1). The exteriors of most elements show a hexagonal close-pack structure, so the gaps between baryons are small. The cubes have gaps between the nucleons that are four times bigger than the gaps in the hexagonal arrangements on the surfaces of the stable isotopes. Baryons are protons and neutrons.

If an isotope candidate starts fusing with a core with a hexagonal close-pack structure, then the surface would have a cubic structure. That would have big gaps on the surface, so it would be vulnerable to fission more than the standard arrangement with a cubic core. Any candidate isotope with a hexagonal core would have no armor on its surface, so they do not exist in nature.

By following the Rules that were shown before Fig. 1, some larger scale structures appear on the surfaces of the nuclei (Fig. 2). Iron has a ring of twelve protons visible on each side of its nucleus. Those structures often are consistent with the properties of the element. In iron, the ferromagnetic property is believed to be caused by the two coaxial rings of protons. A current flows in each ring. If each ring has a current opposite in direction to the current in the other ring, the atom is demagnetized. The rule about protons forming lines of protons is important for creating larger shapes from smaller baryons (protons and neutrons). The rule about a cube being in the cores of most elements results in the excellent stability of most elements. Stability is also provided by the rule for pyramidal stacks of baryons covering the six faces of the cube. Those pyramids often nestle into the low parts of the surface of the cube and the outer surfaces of the pyramids show a hexagonal close-pack structure.

The rule about the top face of the cube being free to build a larger pyramid than the four side faces results in a better armor against a directional flow of colliding particles, as in a nova star explosion.

Figure 2: iron 57 viewed along the axis of the two rings of protons. The view is on the <111> crystallographic axis of the cube. Iron is element 26, with 26 protons.

Models of incremental elements are assembled on foundation element shapes by first finding the deepest pits on the surface and placing any neutrons that are known to exist in an isotope in standard reference books. This static nucleus theory depends on standard scientific knowledge about A and Z [10C], the mass number and the atomic number. Those two integers are already published in many other books and internet websites. For example, see https://periodictable.com/Elements/027/data.html.

Cobalt is element 27. To make the incremental element cobalt from the foundation element iron, the reference books write that A is 59. The mass number indicates that 59 baryons are in cobalt, so first, one neutron is placed on the Fe-57 foundation in the lowest pit, and then one proton is placed in a low pit that follows the rule about protons forming lines of protons. That proton is placed next to the lone proton that is on the axis of a ring of protons (Fig. 3).

Figure 3: cobalt-59 is built on a foundation element

The incremental element copper-63 is non-ferrous. It is based on the iron foundation element. Cu (Fig. 4) is made by adding baryons on top of those existing for the cobalt-59 nucleus that was already discussed. Two neutrons and two protons are added in the lowest pits on cobalt to make the Cu-63 isotope.

Figure 4: Element 29, copper, is not ferromagnetic because one ring is ruined by an added proton

The way that copper is shaped makes sense to people who understand magnetic materials. The three added protons for element 29 make a short circuit from the axis proton to the ring of 12 protons that are on element 26. That is why copper is non-ferrous even though its foundation element, iron is ferromagnetic. This is proof that the theory is correct. This has persuaded the author that the static nucleus theory, using sphere stacking, has the most reasonable Rules for nuclear structure ever proposed. Previous researchers had usually neglected the simple geometric considerations about nuclei with integer numbers of protonic spheres and neutronic spheres.

Chronological order of the discovery events

The discovery of the static nucleus theory took place on May 25, 2017. This chapter gives the sequence of choices that resulted in the correct theory of nuclear structure to be confirmed to be a pyramidal cube. At first, the author realized that other scientists had been unsuccessful when considering the nuclei to be dynamic or chaotic in shape. No application of quarks had given any credible structure to the nucleus of any element. Therefore, the author decided to attempt a static nucleus theory. This would use spherical neutrons and protons (baryons) that are stacked into fixed positions.

Standard crystallography names two compact sphere packing types: the simple cubic lattice and the hexagonal close-pack lattice. The choice was made to begin using a simple cubic lattice at the center of iron, and the hexagonal close-pack would only be attempted at the core if the cubic structure was not reasonable.

Iron was selected as the first element to be evaluated because it has three properties that are obvious and profound. Iron is ferromagnetic to the highest degree of any element. Iron is known to be stable against fission and fusion during any natural process on Earth. Iron sparks when hit. Only cerium also sparks when hit, among all metals. The author speculated that the nucleus of iron has a shape which implies those three properties.

The cube was designed for iron isotope 57. The mass number A is known to commonly be 56 or 57 for iron. It is also well known that Z, the atomic number of iron, is 26. A cube with a stacking of 3x3x3 baryons was chosen because that is 27 baryons for the core of the nucleus. The 27 is less than 56, so it is appropriate. The possible 2x2x2 cube has only 8 baryons and the 4x4x4 has 64, so the choice was made to use 27 protons and neutrons in a simple cubic lattice at the center of the iron nucleus.

Algebra is not as useful as geometry when proposing a static nuclear shape. But some arithmetic proved to be very convincing. Subtracting the 27 baryons of the cube from the A value of 57 results in 30 remaining baryons. Since 30 can be distributed evenly on all 6 faces of a cube, the result was found to be promising. If a pyramid of 5 baryons is nestled onto each face the of cube, then it all adds up neatly.

$57 = 27 + 30 = A$
$30 = 6*5$

The 5 baryons can be a pyramid with 4 in a bottom layer and 1 baryon on a top layer. The bottom layer has a 2x2 square of baryons. In this way, the pyramid has a stable position into which it can be nestled onto a 3x3 square silhouette of a cubic shape. That forms a stable shape, for a stable element. That was a satisfactory beginning to the static nucleus theory, as far as the mass number A is concerned.

The atomic number Z of iron was next considered. By June 5, it turned out to be just as neatly arranged as for A, the mass number. The 27 baryons in the cube could have all protons placed as far away from each other as possible. This would follow the electrostatic repulsion idea for protons which states that protons repel protons. The cube has 8 corners, so 8 protons were placed into the corners of the cube. Arithmetic once again showed a promising result.

$Z = 26$

$26 - 8 = 18$

$18 / 6 = 3$

So, if 3 protons are in each of 6 pyramids, then all 26 protons in iron are accounted for, neatly. The way all of these integers added up to perfection for iron caused the author to consider this to be a successful start to the research. On May 25, 2017, the shape of the iron nucleus was discovered. Its simplicity worried the author, and it was considered that was possibly a flaw. On further consideration, one must wonder why no other scientist had already constructed such a simple model of any elements before. In addition, simplicity is often praised as the mark of excellence. A small degree of certainty began to grow about the correctness of this simple theory.

A plan was made to confirm or disprove the theory of nuclear structure. By late August, 2017 the most important elements had been evaluated and the theory was confirmed. Many criteria were available to show how this idea might be wrong or right. Here is the first set of obvious elements that needed to be evaluated to compare their geometrically implied properties with the known properties. If these four tests were passed, additional test nuclei would be assembled using different criteria.

- Copper must not be ferromagnetic, but Co and Ni must have two rings of protons.
- Gadolinium must be shaped like Fe, but adjacent elements must be different from Fe.

- Manganese must not have two rings like Fe.
- All other elements will be so different from iron that the theory seems to be right.

After days of work all four tests were passed! The author was surprised at how quickly and easily successful checking had been achieved. It is very unlikely that a wrong geometric idea could make gadolinium have the correct shape, like iron, to match the expectations of Ampere that iron has tiny loops of current in a bar magnet. The odds of element 64 being shaped like element 26 must be very low, unless most elements also are shaped like iron. After extensive considerations, it was clear that this is a correct theory of nuclear structure and most elements are not shaped like iron.

A second set of criteria was planned to confirm or refute the static nucleus theory of the pyramidal cube. Since iron has a 3-layer cube, clearly a 2-layer and 4-layer cube must be evaluated. The six faces of those cubes would be covered with baryons that nestle into the pits between the spheres. Those would be analogous to iron in their simple structure. They would then use the Rules to allocate protons into lines of protons. The two new cube sizes would also get protons in places where electrostatic repulsion would be applied. As a result, nitrogen-14 (Fig. 5 and Fig. 6) and promethium 148 (Fig. 7) were assembled using round beads, beans, marbles, chocolates, and bbs.

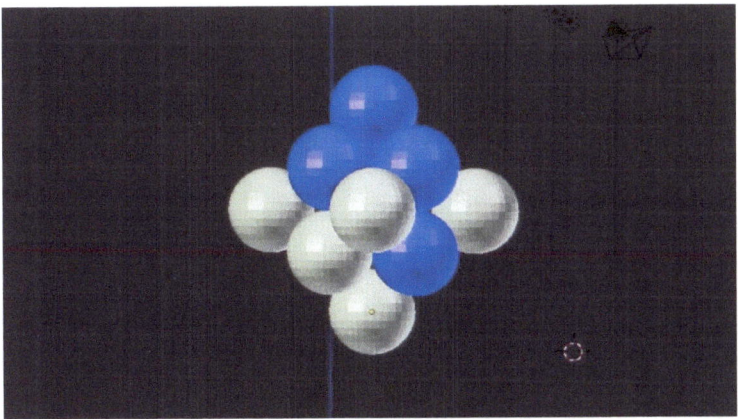

Figure 5: Nitrogen-14 as a pyramidal cube seen on the <100> crystallographic axis. White protons, blue neutrons

Figure 6: Nitrogen-14 seen on the <111> crystallographic axis, implying a triple bond is available

Figure 7: promethium and iron magnified 12 trillion x, next to iron and a cube at 8 trillion x on the right

The three simplest and purest shapes of nuclei are shown above: nitrogen, iron and promethium in Fig. 5, 6 and 7. Each one uses a cube with a different size at its core (Fig. 23). Nitrogen has a cube-2, iron a cube-3, and promethium a cube-4. None of the 118 elements use a cube-5. But it is proposed that when Z=123 and A =305, a cube-5 element can exist. It could be ferromagnetic and stable. Each of those simplest elements has a pyramid on the six faces that is one layer smaller than the cube.

The properties of those three pure shapes are spectacular: nitrogen is for explosives, iron is for magnets, and promethium is made to decay, even without a collision with matter. They teach us about how the shape of the nucleus produces the properties of elements. Nitrogen (Fig. 6) has three protons that are prominently symmetrical, and nitrogen is well known by chemists to have a triple bond. Iron (Fig. 7) already has been described as ferromagnetic, but the nuclear structure teaches us about the reasons why bar magnets can be demagnetized. The two ring currents go opposite directions in a demagnetized iron nucleus. Promethium's structure informs physicists about why it decays so quickly: it (Fig. 7) has a cube-4 so its center point has no matter. Naturally, since carbon also has no matter at the center of its cube-2, one wonders why it is stable. Carbon is stable because there is only one layer of baryons above the vacuum at its center, so the cube baryons have no place to fall. Promethium has two layers of baryons above the vacuum at the center, and it is harder to balance two balls than one.

Highlights of results

After iron was found to convincingly have a simple shape, a model was made using round wooden beads. The 27 beads for the cube were glued together with the white protons on each corner and the dark neutrons in the remaining 19 locations. Six pyramids of 3 protons and 2 neutrons were assembled with the 3 protons in a line. That choice was due to earlier experiments with randomly placing 31 brown beads with 26 white beads in a clear bag. When the spheres were pressed into one big ball of spheres, it became clear that white always touched white beads. It was never found that protons were isolated from other protons by neutrons in the random positions. It is clear that protons must touch protons in stable elements. By extension, if two protons can touch each other, then three or four can also form a line without repulsion driving them apart.

The model was assembled, one pyramid at a time. As two pyramids met, the white proton lines were set to touch adjacent proton lines. See Fig. 1. After 3 pyramids were in place, the author noticed that a ring was formed at a larger scale on one side of the nucleus. More pyramids were attached to the model until only one face of the cube remained uncovered. As the last pyramid was put in place, it was surprising to discover a second ring on the far side of the nucleus. It was like finding a second ring around a moon! As I rotated the model, it was unexpected that two similar rings were seen coming into view. This was not intentional. It was a natural fact. Iron has two coaxial loops of protons and two isolated protons that are also coaxial with the loops. Each loop has 12 protons. All 26 protons are accounted for by adding:

$Z = 26 = 12 + 12 + 2$

The perfection of the shape was shocking. The author became more certain that this was a correct theory of nature. A correct theory will give correct answers quickly and easily for old questions and new questions. An incorrect theory will give correct answers rarely.

When a 4-layer cube design resulted in the element promethium being the most likely element with the 148 baryons, the fast decay rate of Pm was noticed. It seemed like the cube-4 was the cause of the short half-life of that element. Other elements also could be designed with a cube-4 and their half-lives would be considered. Technetium was an obvious candidate for using a cube-4 (Fig. 8 and Fig. 9).

A periodic table of the elements was produced on October 11, 2017 [1C] featuring the gold and iron charge distributions. By that date all elements' shapes had been, drawn in silhouette. Confirmations came over the years with the neon silhouette (Fig. 22 and 37) matching that predicted in France [1A] where Schrödinger's Equation was used in a simulation. Experiments on barium [2A], radon and radium [3A] nuclei showed a pear-shaped nucleus, just as the new periodic table (Fig. 23) had shown. These results are evidence that the new theory was correct. Also, the uranium bimodal mass distribution [5C] of fission fragments fit well with the idea of a 3x3x3 cube at the core of U-234. Fragments after fission had mass numbers that differed by about 27 nucleons. The cube was in one fragment by not the other fission fragment. Since the cube has 27 baryons, the two elements produced when uranium was split apart differed in mass by 27 baryon masses.

New laws of physics are established. Protons make lines of protons. See the non-crossing law. Fig. 39 shows the flat, two-dimensional line of flux that forces iron's protons into lines. Geometric reasoning on page 74 is more important than conforming to the expectations from algebra.

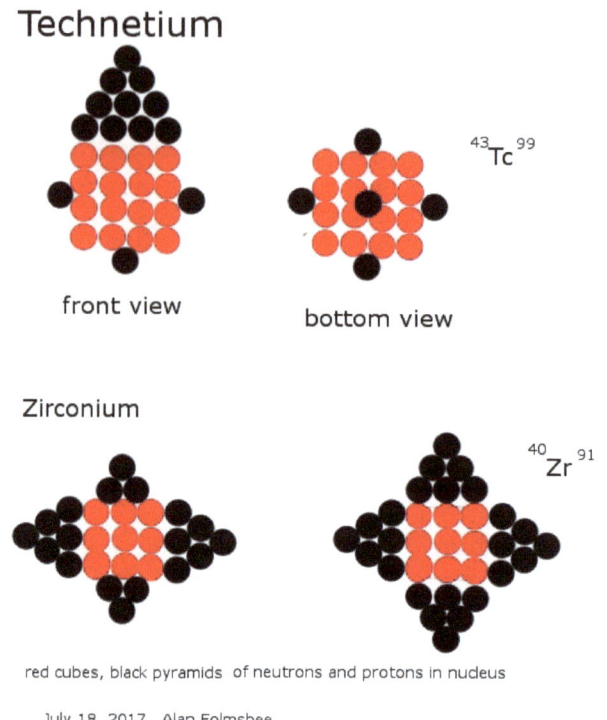

Figure 8: Technetium and zirconium viewed from two angles

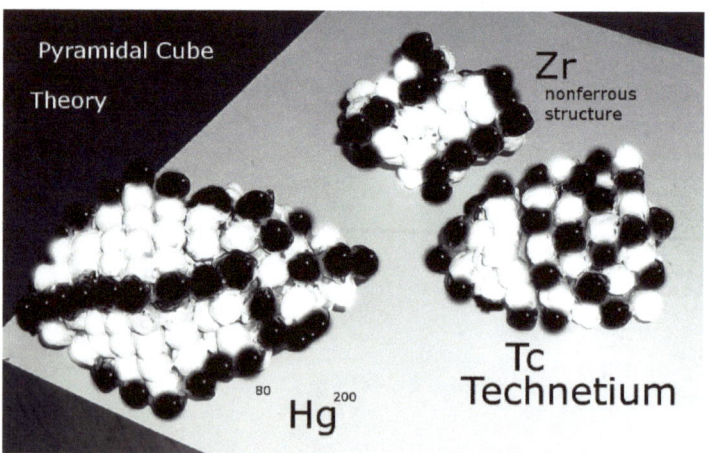

Figure 9: Technetium sculpture near zirconium and mercury

Matter and vacuum

The proton and electron are the most important particles for industry. Neutrons are also useful, but not as much as the proton and electron. The vacuum in outer space is much emptier than in our Earthly vacuum chambers. But even between stars, only about one centimeter separates the particles. In the space between particles is the matter field, to be explained later in this book. Even disregarding gravity, the vacuum is filled with dimensions that connect electrons to protons. Those dimensions are what force protons to form lines of protons in the nucleus. Invisible in the vacuum, the matter field's dimensional lines and velocities will be defined later in this book.

In high school, we are taught that opposite charges attract and that protons repel protons in every direction. But charge is not so much like gravitating matter. The charge concept has only two real aspects. Charge is a category and charge is compared to mass in a ratio. For example, the formulas of physics use Coulombs that are usually cancelled, leaving a force or a velocity or acceleration as the important result. Look at the old Coulomb force formula. All Coulombs get cancelled. Look at any physics equations in your text books and you will find Coulombs get cancelled out. The second aspect of charge is more profound, where a "charge to mass ratio" is used in formulas. In an abstract sense, mass is an area and charge is an area. That is why their ratio is meaningful: area divided by area is dimensionless, so the result is a magnitude.

The point is that protons form lines of protons in the nucleus, not because Coulomb repulsion has some new complication, but because electrons are associated with protons, and the electrons force the protons to be excluded from certain positions. (See the chapter on electrons). Near a nucleus, the vacuum is densely populated by matter fields. Particles behave differently near a nucleus than they do when separated by farther than an atomic radius. Nuclear structure is calculated by using spherical protons and neutrons. The radius of a proton is about 0.85 femtometers. Please do not consider that radius as where something begins, but where all measurements end. The proton is a cavity in our measurable universe, not a lump of solid stuff. It is a sphere with a perfection of form that we can use to model the stacking of spheres to be profoundly accurate. Negative space is inside the proton. At the center is a mirror for four dimensions for gravity (x, y, z, t). There is traction [1C] at the mirror so that electromagnetism emerges as four more dimensions (hx, hy, hz, and th). The dimension hz feeds any neutrons that are close by. For the hydrogen isotope 1, there is no neutron, so hz leaks out into the universe to cause expansion.

Matter is not just mass. Matter causes gravity because protons have a volume. Electrons do not have a volume, so they do not cause gravity. Masses are attracted by gravity, but gravity is not caused by mass. It is unfortunate that the letter m starts both words. Some science popularizers misinform children that e equals mc squared uses matter instead of mass in the formula. In that way, people mistakenly expect that matter is destroyed during nuclear fission. No. Mass gets reduced but matter has the same number of baryons before and after fission. No matter is destroyed.

The vacuum allows radio waves to propagate only if matter is present in the vacuum. There is always matter present, even between galaxies. Light is different from radio waves. Radio is not photons, but since there is no perfect vacuum, they are hard to distinguish.

The element iron has matter at its core. Promethium has a vacuum at its center (Fig. 10). Both elements have simple shapes, but one is stable and one decays away to nothing after a few thousand years. Promethium has a four-layer cubic lattice, so there is nothing in the middle. Iron has an odd number of layers in its cube, so a neutron is in the middle position and the rest of iron is on top of that solid base. If the theory is correct, about time being emitted from each neutron and proton, and about the inability of time to pass through those baryons, then the difference between odd and even cubes is most stark in the center point of those two symmetrical elements. Let's look at the silhouettes of Fe and Pm.

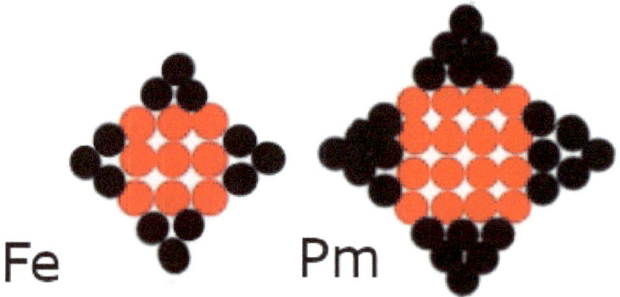

Figure 10: matter at the center of iron, vacuum at the center of promethium

The matter field is believed to exist in the center of promethium, in a vacuum. This is the smallest vacuum chamber ever discovered. The instability of Pm-148 is similar to the instability one experiences if a stack of two basketballs is balanced on a spike (Fig. 11). The stability of Fe-57 is similar to that experience by a person who tries to balance one basketball on another basketball. It is difficult using the spike and easy to balance one ball on a ball. It may be impossible to balance three basketballs in a vertical stack on top of a spike. These three experiments you can try at home.

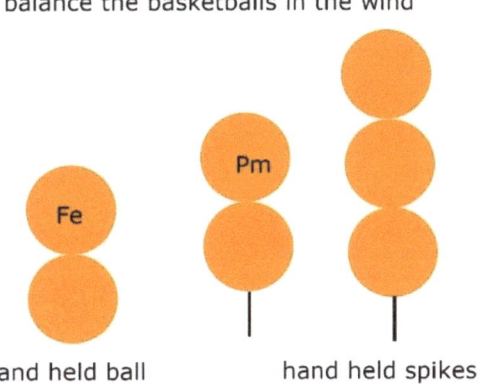

Figure 11: Stable and unstable sphere stacking thought experiments

If you have some basketballs, you can try the experiments. There can be calm air or a windy condition is also appropriate to represent a matter field. A fan can blow a breeze upwards on the basketballs to attempt to stabilize the stacked spheres. The spikes represent the difficult from having a vacuum in the center of an isotope. Iron is illustrated by holding one ball in the hands with skill and balancing a second ball on top of the first ball. That is easy to balance. In the middle of the picture in Fig. 11, is the promethium stack, where the vacuum is at the bottom of the stacks, represented by a spike that is hand-held. The spike holds a first ball and a second ball must be balanced on that stack. It is difficult, but maybe a skilled basketball player or magician could do it. If a fan blows a vertical wind up past the stack it might be easier. That wind represents a matter field, to be defined in the chapter on electrons. On the right side of the picture is a stack of three balls on a spike. It might be impossible for the most skilled person to balance that stack of spheres. But with a wind blowing upwards past the stack, it would be easier. The matter field is like a wind or a liquid current: it can stabilize a stack of spheres or it can blow it apart.

Iron versus copper

Copper is element 29 and it has three more protons than iron. Elements 27 and 28 are cobalt and nickel, two ferromagnetic elements. When one proton is added to nickel, the resulting element is nonferrous and diamagnetic. When copper is placed near a nickel bar magnet, it is repelled with a tiny force, regardless of which pole is being applied to it.

Why does adding one proton change the magnetic function of copper so completely? For years, scientists have wondered about that. Now, a clear answer has been discovered. The Static Nucleus Theory of the Pyramidal Cube provides the answer. The added proton ruins one of the rings of protons so that only one ring exists for copper. A short circuit between the axis proton and a ring exists for copper. To completely describe this discovery about copper, the following text is copied from [1B] The Journal of Nuclear Physics, where this paper was published in March, 2019.

"Magnetism from iron's nuclear structure"
Author: Alan C. Folmsbee
Affiliation: Independent Researcher
Address: Haiku Road, Haiku, HI USA 96708
Date: March, 2019 Issue of The Journal of Nuclear Physics (JNP) website

Abstract
In this proposed theory, the nucleus of the element iron has a shape that causes ferromagnetism. That shape also causes the stability that is the best of all elements. The protons in iron make loop shapes around the exterior of the nucleus. The loops are coaxial. The iron nucleus has a cube of protons and neutrons at its core. The faces of the cube are covered by pyramids of protons and neutrons. All ferromagnetic elements have the coaxial loop structure like Fe. No nonferrous elements have that shape, within tolerances. Most of the properties of the elements are related to the geometries of the nuclei. A new periodic table articulates the silhouettes of elements that were compared with the structure of the iron nucleus to be certain that all elements are consistent with the Pyramidal Cube Theory.

Geometric Theory of the Nucleus

The radius of the proton has been measured in the past to be about 0.9fm. This is interpreted as the radius of a spheroidal baryon. This spherical model will be used, without any quark substructure considered. Protons must touch protons. That geometric fact is easy to demonstrate for yourself. Use 57 spheres of two colors to put in a clear bag. For the common isotope Fe 57, this is 26 dark spheres and 31 light spheres to represent protons and neutrons. When the spheres are pressed together, a candidate nuclear mock-up can be produced by hand. The random locations of the protons are seen to make protons touch protons in more than half of the cases. Fig. 12 shows an example of this using a corner of a box with 57 spheres in random positions. It is impossible for neutrons to insulate each proton from all 25 other protons unless a long line of spheres is formed. Even a deliberate positioning of the protons can only result in protons touching one or two protons, when trying to approximate a spherical nucleus. That must be stable. The conclusion is unavoidable: in the iron nucleus made by sphere stacking, it is required that protons touch other protons. That is a stable arrangement that can be used during the construction of this mock-up. This is a profound fact that needs to be emphasized: protons touch protons in stable nuclei. Neutrons are not needed to be positioned between protons to isolate them.

This theory proposes that the proton positions are static. The neutrons have stationary positions within a lattice called the pyramidal cube. Protons have permanent positions within a pyramidal cube of baryons. The nucleus is a placid place. When several protons form a line of protons, that is stable in the nucleus of an element. That line is accompanied by a line of neutrons in many elements. The nuclei are not random mixtures of moving spheres with a changing shape. Nuclei are made using static lines of protons in the pyramids. This arrangement is responsible for the A/Z ratio increasing with Z. It is expected from this theory that the ratio A/Z is between 3/3 and 8/3 for all elements. The heaviest element designed with this pyramidal cube theory has a A/Z ratio of 799/298, which is close to 8/3. In other words, the number of neutrons in a nucleus asymptotically approaches 1 and 2/3 the number of protons. The cube in the center of the iron nucleus has no protons at the center, as designed for Fig. 13.

5 Sphere Stacking Rules Used for Iron

A cube of baryons is designed at the center of the nucleus.

Protons in the cube are far from each other.

Outside the cube, protons tend to form lines of protons, seeded from the cube.

A pyramid grows to completion on each face of the cube.

Protons are sparse at the center of the nucleus and densely allocated near the tips of the pyramids.

Figure 12: Gray neutrons and red protons in mock-up

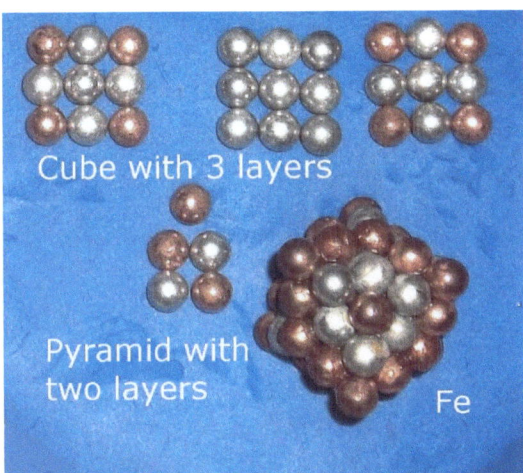

Figure 13: Articulated nucleons for the cube and pyramid

The Cube of Baryons at the Center of Each Nucleus

The iron nucleus has a central part that the author hypothesizes to be a cube. That core cube has 27 protons and neutrons. It is like a crystal lattice using a cubic stacking in Fig. 13. The cube has six faces. A cube of protons and neutrons with a size of 3x3x3 has 27 baryons, so that is appropriate for iron 57. If a cube of 2x2x2 were used, only eight out of the 57 baryons would be accounted for. A 4x4x4 cube would have 64 baryons, exceeding the limited mass number A for Fe 57. When a nucleus is formed, baryons become nestled into the low areas between the cube spheres. The cubic lattice meets a different lattice type of the pyramid in Fig. 14. This combination can be called a tetra-hexahedron [4A]. Another phrase for this lattice type is Face-Armored Cubic.

The simple cubic structure is rare in solid elements. Polonium has that crystal lattice. More common is the face centered cubic lattice in an array of atoms. For the iron nucleus, the simple cubic core is always stabilized by pyramids on the faces. This new lattice type is called face armored cubic. For the cubes of heavy elements, the protons are present near the center. There can be alternating neutrons and protons in the 3D geometric models that form a checkerboard matrix of 3x3x3 protons and neutrons.

The six pyramids must contain N baryons, with N being the difference between the mass number and the count of baryons in a cube:

$N = 57 - 27 = 30$ baryons outside of the cube

N baryons must be on six faces of the cube. Each face needs M baryons. Let M be the baryon count piled on each face of the cube:

$M = 30/6 = 5$ protons and neutrons per face of the cube

Five baryons can make a pyramid with two layers. There is a base layer with two protons and two neutrons and there is a capstone always made of a proton. In Fig. 13 through Fig. 15 are shown the three layers of the cube for element 26: Fe 57.

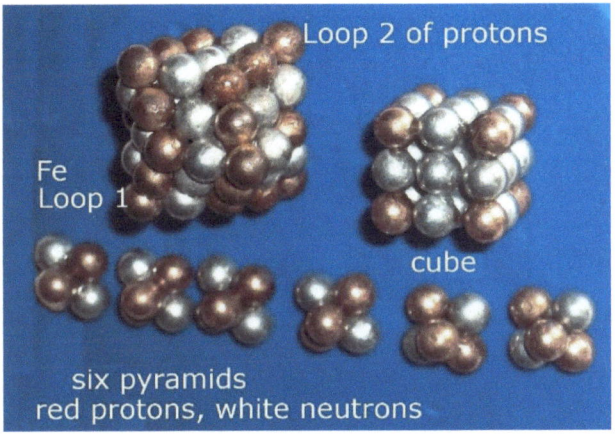

Figure 14: Protons modeled with copper bbs, neutrons zinc

Eight protons are positioned at the corners of the cube. This was a design decision meant to mimic the situation measured on common electrical apparatus, where the excess charges tend to be at the exterior surface of the matter. It is true that a cube has eight extreme points. That is why an attempted shape design was made starting with eight protons in the cube-3. The remaining eighteen protons are divided among the six pyramids. That means three protons are in a pyramid of five baryons. The protons must touch protons. A symmetric line of protons is formed in the pyramid, as a shape preferred over the alternative, non-symmetric allocation of protons next to neutrons. That alternative proton allocation is not shown. There are choices in the rotation of each pyramid as it is placed on a cube. The choice was made to line up lines of protons with other lines of protons. If a pyramid is rotated 90 degrees away from that line-up, a T shaped intersection of proton lines will exist. The choice was made to consider the T line-up to be disfavored by proton interactions, compared to the situation where a proton is placed in a line with other protons.

Figure 15: Dark protons, light neutrons

     Iron was chosen to be the best candidate for trying to relate the element's properties to the shape of its nucleus. Iron's ferromagnetism, stability, and one other property are spectacular, compared to other elements. Iron sparks when struck and the only other element that does that is cerium. It was speculated that the shape of the iron nucleus would have some relationship with magnetic phenomena and stability. That was found to be a realistic concept of the nucleus in the geometric work that followed. The nuclear stability is due to the shape of the pyramids. Iron 56 and isotope 57 were both modeled and both are stable against nuclear decay. Iron 57 is less common, but it provides the best pyramidal cube paragon. Other exemplary elements are nitrogen and promethium, where the cube-K has six pyramids of (K − 1) layers. That is seen in Fig. 23, the periodic table.

Figure 16: Five pyramids on a cube

Figure 17: Dark protons, light neutrons

The cube at the center of a nucleus has six faces that are covered by piles of neutrons and protons. Iron has a cube with eight protons and nineteen neutrons for the 3x3x3 baryons. That is called the cube-3. See Fig. 16 to see one pyramid removed from the nucleus to reveal the cube. It is proposed that all cube-3 elements come from cube-4 elements that failed to obtain pyramids on all sides. Only Technetium and three other elements have a cube-4 in their centers with six completed pyramids to armor the candidate nucleus during creation.

The pyramids are formed in two common fusion events. First, an omni-directional heat causes abundant collisions. Second, a unidirectional blast causes collisions during a week-long flood of candidate fragments going at the target nucleus. That forms heavy, elongated elements and a more unbalanced top pyramid shape compared to the bottom pyramid. The heat related fusions create pyramids that have less elongation of the nucleus. Those elements are seen in a proposed periodic table in Fig. 23.

The proposed theory of the creation of the elements includes the reason why technetium is so different from similarly lightweight elements. The sequence involves the triple alpha process to make the carbon cube-2 core. Then a cube-4 core is made from eight cube-2 nuclei. Any cube-4 candidate element that fails to cover its six faces with pyramids will decay into a cube-3 element. A cube-3 that encounters a lithium rich environment will make iron using six Li 5 isotope pyramids.

Observing the Geometry of the Mock-up

This theory of the shape of the iron nucleus uses geometry. That is one choice between the use of the algebraic versus geometric schools of mathematics. Using the pure mathematics of geometry, a person holds the three-dimensional mock-up in the hands, rotates it, looks at it from many angles. Insights are obtained by the researcher in ways that algebra does not provide. Fig. 14 shows the iron nucleus from a perspective that reveals both loops of protons modeled as copper spheres. The iron geometric model was handled by the observer to evaluate any insights that are available. The two loops of protons are visible and they surprised the author when the mock-up was first assembled on May 25, 2017. If algebra is used for modeling a neon nucleus, one technique in [1A] gives a probability distribution for the positions of the baryons in a nucleus of neon.

The Stability of Iron

The nucleus of iron is very stable because its shape is optimally allocated to have no vulnerable areas, compared to all other pyramidal cube shapes. Iron is stable against decay and it is difficult to fission iron or fuse it. Many shapes were predicted for the elements with known properties like Z and A, atomic number and mass number. Those shapes were compared to the shape of iron to see if any other element has a better shape for stability. The stability of iron is explained geometrically, without needing sub-particles below protons. This geometric evaluation features the gaps between the pyramid nucleons being more invulnerable to incoming matter than are elements with protruding ledges, like polonium.

The pyramids are narrower than the cube for the iron nucleus, as in Fig. 15. The two-layer pyramids on the three-layer cube makes a clogging concept seem realistic. The baryons are draining a fluid and the gaps between spheres will be filled in with incoming spheres. This is not a geometry in which a force at the center of the nucleus pulls on the nucleons and the force passes through the spheres. The force is centered on each nucleon. The spherical nucleons block the flow of a fluid that goes around the outsides of the protons and neutrons. This makes higher time derivatives become important. Those derivatives are proposed to allow gravity become the strong nuclear force. Iron is stable because the pyramids stack with a structure that plugs the gaps between cube baryons. The cube would be vulnerable if the faces were not covered well. Cubes have gaps about four times as big as gap areas on pyramids. Gaps between spherical nucleons are where a strong nuclear force is available in this hydrodynamic model. Larger pyramids than those on iron would stack without nestling the spheres as they stack. That is less stable than nestled spheres of a pyramid. An element with a four-layer pyramid would overlap the cube and that overhang would be vulnerable in a collision. The shape of iron armors the nucleus against additional fusions. All of the gaps on iron's surface are small gaps, compared to those in a cube.

Recognizing the Cause of Ferromagnetism

The shape of the iron nucleus was discovered, not designed. The author did not plan or expect two loops of protons to be in the shape of a coaxial connector. The author designed a cube to be at the core of iron's nucleus. The two loops were observed after the mock-up was assembled as in Fig. 17. There are few choices that can produce a non-coaxial shape. A logical assembly of lines meeting lines is the choice that gave the insightful results. That figure helps the reader to appreciate the scale model's sculpted shape with a loop of protons in iron. This view is on the <111> crystallographic plane, relative to the cube. This is looking straight into the coaxial loop structure with a magnification of five trillion for the mock-up using 4.5mm metal bbs. The two loops were recognized using electrical engineering judgement to be the cause of ferromagnetism.

The goal of this research was initially to see if sphere stacking could explain ferromagnetism and stability. When the mock-up observations yielded surprisingly good results, a new concept of magnetism was immediately proposed. A loop of twelve protons in iron is polarizing and combining their fields to emerge from a coaxial nucleus. That is the magnetic flux that pairs with twelve distant electrons. Those electrons can be making eddy currents in a remote bar magnet. According to this proposed theory, Ampere was right: there are loops of currents in the nucleus and that can cause a looping of twelve electrons that are far from the iron magnet.

Iron is a pyramidal cube,
a tetrahexahedron

Figure 18: Fe has coaxial loops of protons

The Curie temperature can be explained by using the two loop currents going the same way or with random directions. In that theory, the electron temperature will affect the proton loop current direction. At low temperatures the two loop currents can oppose or go the same way in a nucleus. At high temperatures the two loop currents will have randomized directions due to the electron interactions with the nucleus.

The fact that iron has loops of current at a small scale was first proposed by Andre-Marie Ampere in 1820. Today, loops are common in electric devices like transformers, so it is easy to recognize the relationship between a shape and a magnetic phenomenon. In transformers, a primary loop and a secondary loop of wire provide a shape that is magnetically significant. That is true for the iron nucleus, where the geometric reasoning of the pyramidal-cube Rules produces two loops of protons, as in Fig. 18. This view is on the <100> crystallographic axis of the cube. The current is in the loop even though the twelve protons are not moving. A dimensional flow is making a current, not a relative motion of particles. That loop creates a harmony and collective behavior of the flux which emanates from the iron nucleus. The author suggests that in a bar magnet, the primary loop of 12 protons sends its flux out through the center of the secondary loop and it goes to a remote bar magnet with electrons that are paired with protons in the first bar magnet of iron. Flux bundling allows magnetization to occur from an outside flux. Another example of magnetic effects being well known for two loops is the example of two antenna.

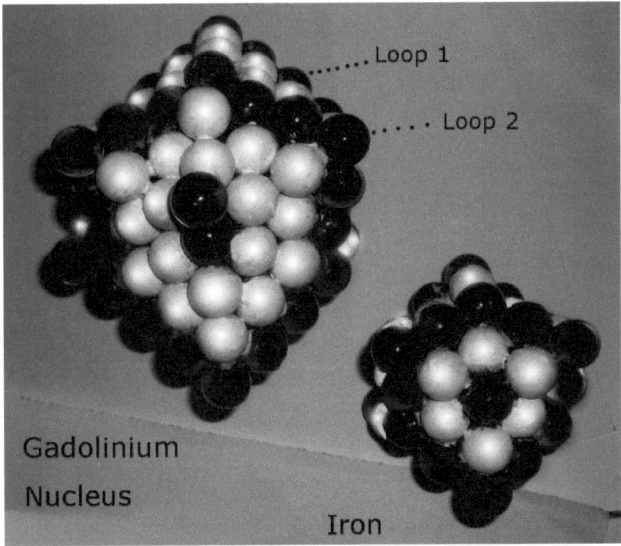

Figure 19: Coaxial shape for ferromagnetism

Tolerances for Geometric Irregularities

The elements Fe, Co, Ni, and Gd all have the two loops. Gadolinium has provided strong confirmation of this static sphere stacking theory. See Fig. 19 to compare element Z=26 Fe and element Z=64 Gd. The main difference is Gd has twice as much neutron isolation between loops and Gd has eighteen protons in each loop instead of twelve. The degree of perfection of Gd is less than for Fe because the six pyramids are not all alike for Gd. The top and bottom pyramids are bigger than the side pyramids, unlike iron. At the coaxial centerline, there are some protons out of the axial line. But that misalignment is in a plane that intersects the tips of the two large pyramids. Those two imperfections seem to cancel each other to allow gadolinium to be ferromagnetic.

Iron has the perfect shape of a tetra-hexahedron. Cobalt is like that but it has an extra proton near a loop in Fig. 20, so cobalt is not as magnetic as iron. Manganese and copper are not ferromagnetic

Figure 20: Nonferrous element loop shapes are not tolerable to have a superconducting current

and the figure shows that copper has a proton that ruins the loop shape. Most elements are not ferromagnetic. To prove that the pyramidal cube theory is correct, it is necessary to evaluate all elements to find out if some elements other than Fe, Co, Ni, and Gd have the coaxial shape of two loops.

Gadolinium is an incremental element based on cerium. The elements Fe and Ce will spark when hit with a hard edge. The shape of Ce is like Fe, but the top and bottom pyramids are two layers larger.

Carbon has a 2x2x2 cube with two protons and six neutrons. Iron has a 3x3x3 cube with eight protons and nineteen neutrons. Tungsten has a 3x3x3 cube with a checkerboard pattern of protons in the neutron matrix of the cube. Fig. 21 shows the elements Tc, Hg, and Zr. Those structures were evaluated and it was discovered that none of the non-ferrous elements have the two coaxial loops of protons. Zr has two loops that are not coaxial. Tc is modeled with a checkerboard pattern visible in the model, since the cube 4x4x4 has only a 1-layer pyramid on five faces. That sparsity of protons covering the faces lets the cube be seen in its raw structure.

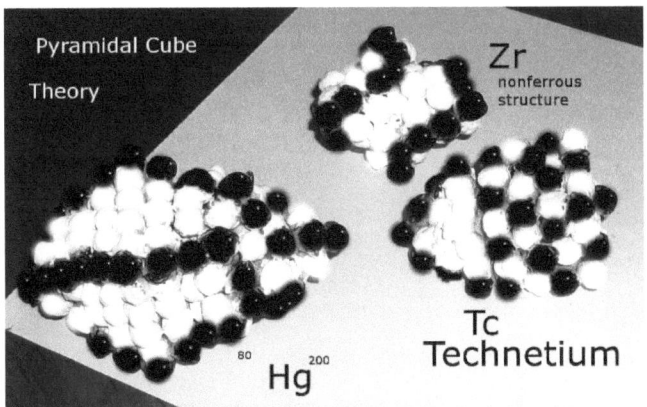

Figure 21: Pear shaped Hg near Zr and Tc

Confirmation using Neon, Barium, and Radon
    The confirmation of this geometric theory of the shape of the iron nucleus is provided using algebra and experiments of other researchers. The algebra of Schrödinger's Equation was used in [1A] for neon. In that paper, a color image is provided for the silhouette of neon's probability that a nucleon is in a position. The shape of that nucleus is the same as the shape of neon in Fig. 22 where the neon mock-up is using yellow spheres for protons and gray spheres for neutrons. This shape of neon as a pyramidal cube matches the silhouette provided by other people in [1A]. Experiments for barium and radon show a pear shape for those nuclei [2A, 3A]. Those papers gives a color image of a blurred pear shape. That shape is like in the periodic table of nuclear shapes for barium and cesium. The pear shape is also seen in Fig. 21 for the mock-up of the mercury nucleus. The uranium model in Fig. 22 also shows the pear shape, due to a six-level pyramid at one end of the nucleus. Iron did not have reference information available in the ways neon and barium were available from independent people who showed projections of the shape of a nucleus. Iron has been expected to have an almost spherical shape. The new theory of iron's shape is confirmed due to several factors. The neon shape in [1A] has the right proportions to have a cube 2x2x2 and that is indirect confirmation of iron as a pyramidal cube 3x3x3 nucleus.

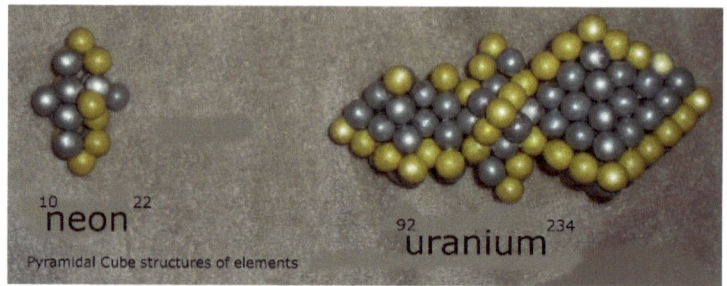

Figure 22: Neon and Uranium mock-ups

Ramifications

The static nucleus theory has protons and neutrons in fixed positions for each element. This leads the way for future research on chemistry and magnetics. Chemical property predictions are realistic from this theory. The 2s orbital concept for light elements is consistent with this model because the cube-2 has 2 protons. But from iron upwards, elements could be modeled with a new 8S nomenclature attempt because the cube has 8 protons in iron. See Fig. 23 for the Periodic Table of the Elements. The nuclear structures of all 118 elements are shown in silhouette. The red highlights the cubes of baryons and the black identifies the pyramids of baryons. The blue shows the protons of incremental elements that were added on the exteriors of foundation elements. Inert gases can be supplanted as the basis for describing heavier elements. Instead, the foundation elements will provide a more articulate way to name orbitals of elements. Paramagnetic and diamagnetic elements can be matched with the small loops seen in some elements. The cross-sectional areas of these pyramidal cubes can be used to calculate the mass. A new stable element number 123, isotope 305 is proposed. It is expected to be ferromagnetic. The strong nuclear force can be understood as being made by gravity near the shapes in the nuclear sphere stacking that provide curved gaps for producing higher derivatives of time. The space sinking into the nucleus and the time growing out of the nucleus are constrained by the tightly curved radius of each nucleon. This curved spacetime is what makes the strong nuclear force out of gravity. The time is thickened where it passes through the gaps, so higher time derivatives are employed to go beyond acceleration. This shows how the strong nuclear force is gravity.

The pyramids on iron have three protons, like lithium. It is proposed that lithium abundance is suppressed in the Sun because iron and many elements fuse lithium to be a pyramid of Li 5.

The Periodic Table of the Elements

The silhouettes of all 118 elements are shown in the Periodic Table of the Elements in Fig. 23. You can also get more detailed sections of this table on the website on the internet called pyramidalcube.blogspot.com. On June 19, 2021 the best version became available there on a link, in the reference section [20C]. The red circles represent baryons (protons and neutrons) that are stacked into the simple cubic lattice. The black circles are baryons in pyramids that cover the six faces of the cube. Since this shows silhouettes, the front pyramid is not shown so that the internal structure clear. The pyramid on the back side is also not visible here. The left and right sides of the cubes are shown to be covered by identical pyramids. All four side faces of the cube are covered by the same sized pyramids to armor the cube. The top face and bottom face of the cubes can have different sized pyramids covering them.

In general, the periodic table here shows the shapes needed for survival during eternity. The eight groups on the right side of this table are mostly diamagnetic, but period 7 is not fully defined by data. A space was put in the table to highlight the right eight columns. Some references say plutonium is paramagnetic and some do not know. This version of the periodic table does not separate the lanthanides or actinides. This is a way to be more articulate about nuclear structure. For example, Tc is in the same group (column) as Pm. A space is put into this table so those cube-4 elements and cube-2 elements are on the left and cube-3 elements are on the right side of that space between the seventh column and the eighth column. The old nomenclature about groups for chemical purposes is not obeyed for the nuclear structures. This discrepancy will need more explanations in the future. In this table, 32 groups and 7 periods can be defined. In standard periodic tables, only 18 groups are named.

The older periodic tables of the elements were made to highlight the chemical properties of groups being similar. The nuclear structure was of no concern in the olden days before nuclear structure was clearly defined. The lanthanides and actinides were separated from the other elements so the printers could fit the table on pieces of paper or wood that could be manufactured without the use of digital light emitting diodes or liquid crystals. Today, the table can be displayed on electronically powered flat screen display inventions that Dmitri Mendeleev did not have available in 1871. Unfortunately, if one wants to compare technetium and promethium on the old table, they were scattered into positions that destroyed any insights into nuclear decay modes.

More information about all 118 chemical elements' nuclei is available in [22C]. That long reference book gives close-up views of the Periodic Table's cross-sectional shapes of nuclei.

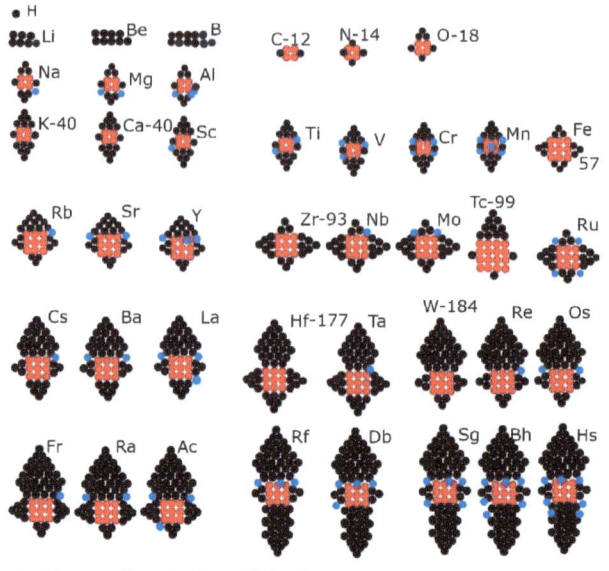

Red baryons in a simple cubic lattice. Black baryons cover 6 faces of

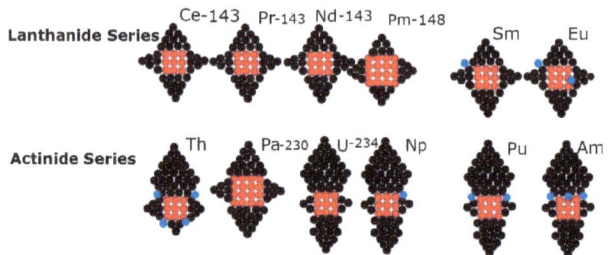

Periodic Table of

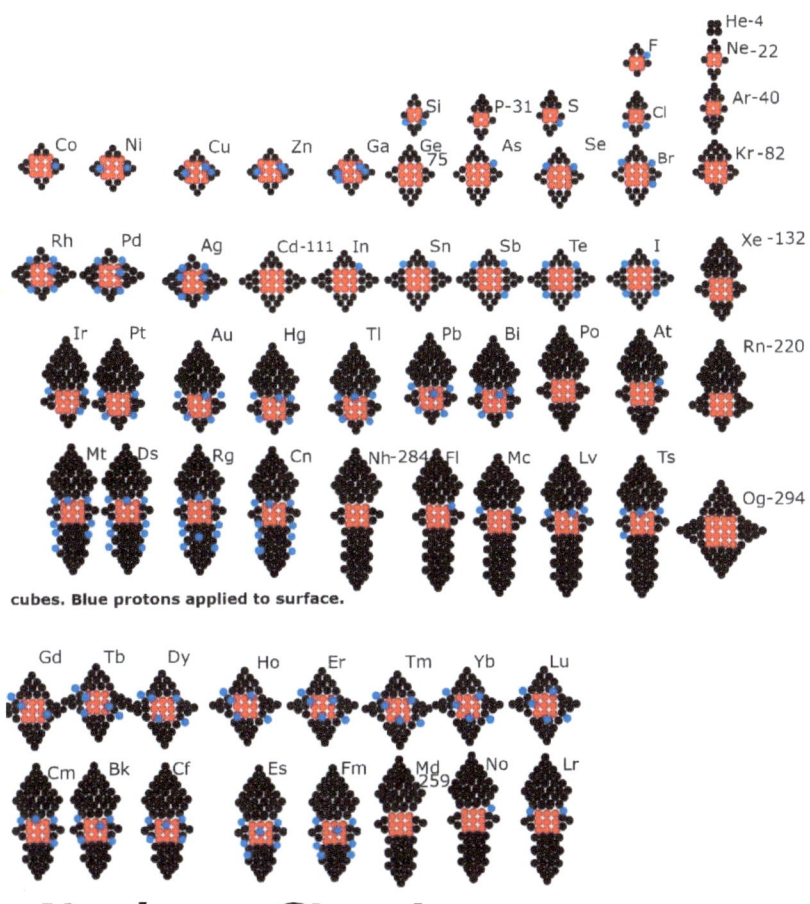

Nuclear Structure

Figure 23 on two pages, left and right pages for The Periodic Table of Nuclear Structure

Conclusion

Using engineering judgment, the author asserts that this theory of iron magnets is more realistic than any previous understanding of the shape of the nucleus. This certainty about iron allows inductive logic to be used to define the shapes of all 118 elements. The evidence is so abundant that it is important for scientists to someday have full confidence that this pyramidal cube theory is the correct theory for the structure of the nucleus. That confidence should be based on observing how the properties of element 26 match the geometries of its nucleus. This synthesis of the measurable forces from iron and the theoretical crystalline shapes of one nucleus has provided the geometric reasoning for expecting all elements to follow the same tendencies during the creation fusion. This sphere stacking approach is better than the algebra of Schrödinger's heirs for understanding how the shape of a nucleus brings insights into the nature of the magnetic flux from iron that is paired with remote electrons.

Acknowledgments:

The author acknowledges the work of many scientists who used the mass spectrometer of Francis Aston to provide the mass numbers of all of the elements. Those facts enabled the geometric reasoning for the pyramidal cube theory of the shape of the nucleus.

References Section A

1A "How atomic nuclei cluster", J. P. Ebran, Elias Khan, T. Niksi and D. Vretenar, Nature, July 19, 2012

2A "Direct Evidence of Octupole Deformation in Neutron-Rich 144 Barium" B. Bucher, Phys. Rev. Lett. 116, March, 2016

3A "Studies of nuclear pear-shapes using accelerated radioactive beams" Gaffney, Butler et al. May, 2013, Nature

4A "The Sciences of Chemistry, Minerology, and Geology", J. G. Heck and S. F. Baird, 2013, ISBN-13: 978-1236959133

Carbon is unique

The elements of DNA are known to include carbon, nitrogen, oxygen, phosphorus, and hydrogen. But carbon is usually named as most important. It is so common that it is not even labeled where it exists in a DNA chemical structural diagram. Every other element is labeled explicitly. What is special about carbon? In the molecules, carbon is pictured as connecting together other elements in a chain or in a ring of atoms. The static nucleus theory has a way to explain this chaining and also the flexibility of the fibers in living plants and animals. In Fig. 24 are shown the six protons and six neutrons of carbon-12.

Figure 24: Carbon 12 with green protons and light neutrons, the cube is visible.

Notice that carbon has three protons in a crooked line on the left side and three more protons on the right in a second crooked line. A group of three protons defines a plane (Fig. 25). The two groups are symmetric to each other, relative to the center of the nucleus. The cube has two layers, with the top layer easily seen with one proton. The bottom layer of the cube also has one proton and three neutrons. That means there are two protons in the innermost zone of the nucleus. That is similar to how electrons are said to have two electrons in the inner shell. The s orbitals of carbon and all light elements have two electrons because the nucleus has two protons at its core (Fig. 26).

Graphite is made from planes of crystals that slip over each other. That material has that property because the nucleus has two planes of protons that are parallel and symmetrical. No other element has two parallel planes of protons that are symmetrical. That is what makes carbon unique. It can equally treat two outside atoms in a chain. The two bonds from carbon to two other atoms do not touch each other, so it allows flexibility. Only the nucleus holds the two bonds together, without needing the bonds to touch each other. Carbon treats two adjacent atoms equally because of the symmetry.

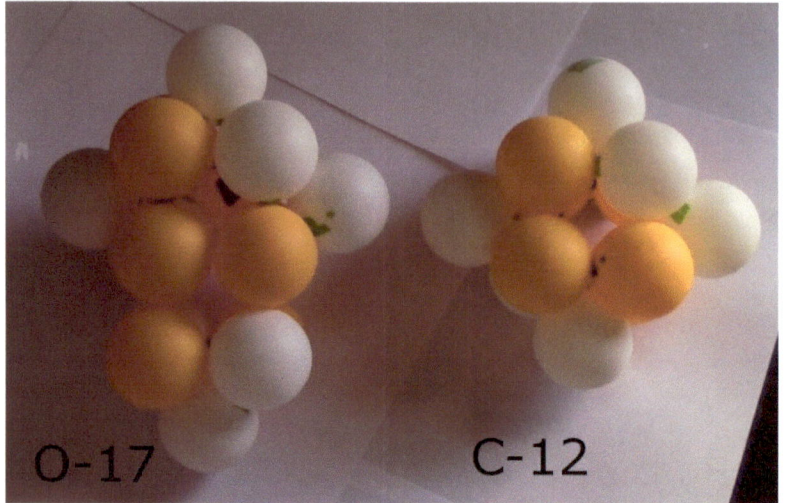

Figure 25: Top view of carbon-12 nucleus, green protons, clear neutrons

Figure 26: carbon has 12 baryons, green protons, clear neutrons

The symmetry of carbon-12 makes it invisible to MRI, magnetic resonance imaging. That is why carbon-13 is needed to measure the positions of carbon in any molecules. Only 1% of carbon atoms are carbon-13. Technicians for MRI need to obtain enriched materials with more abundant C-13 isotopes present. The pyramidal cube theory explains why C-13 is measurable and C-12 is not. When a magnetic B field is applied to carbon-12, the symmetry makes it unable to be vibrated due to equal forces and masses being present. The added neutron on carbon-13, makes the nucleus become unsymmetrical in its moment of inertia. It is lop-sided, so the B field can unevenly affect the nucleus to make it resonate. That effect is similar to the way a round ball will not oscillate when thrown, but a dart will wobble. That is evidence that the static nucleus theory is correct to use the pyramidal cube structure for the carbon nucleus. The proof is adding up to bring certainty.

Cerium, Neodymium, and Gadolinium

Cerium is a foundation element with 58 protons (Fig. 27). Neodymium has two more protons and gadolinium has six more. The periodic table is provided in Fig.23. In that table, Ce is seen to have a silhouette with a cube-3 that has the top and bottom faces covered by 4-layer pyramids (pyramid-4). The four sides have a pyramid-3 (Fig. 28) on each face of the cube. It is similar to iron and it sparks when hit.

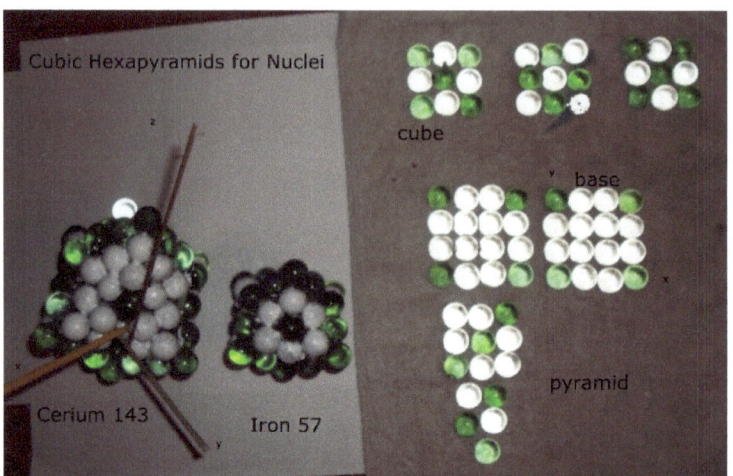

Figure 27: cerium built from two 4-layer pyramids, a cube, and four 3-layer pyramids. Green protons and white neutrons.

Figure 28: cerium has a cube-3, two pyramid-4s, and four pyramids with 3 layers

To make a neodymium model, add some neutrons and two protons. In Fig. 29, the red are baryons in a cube, the black are baryons in a pyramidal stack, and the blue are protons added on a foundation element.

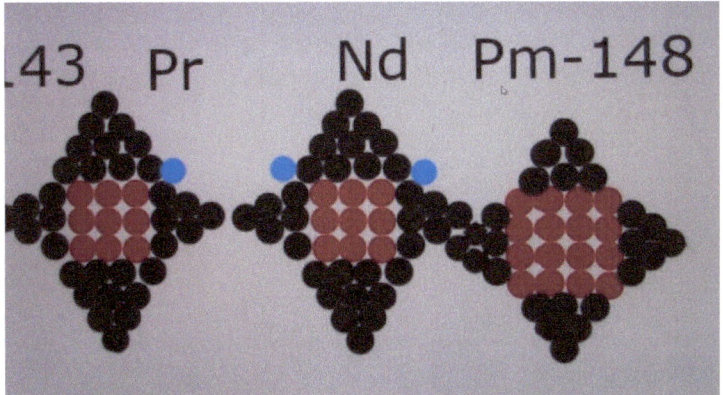
Figure 29: neodymium silhouette next to praseodymium and promethium plans

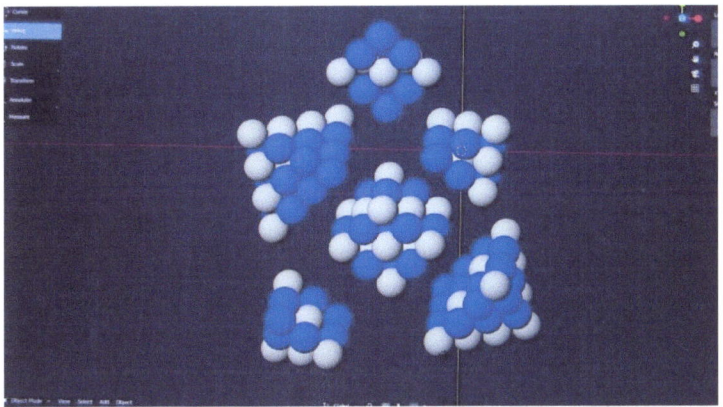
Figure 30: for a 3D printer, components of neodymium and cerium

Digital files for 3D printers were made for cerium (Fig. 30, 31) and neodymium using the Blender software. You can download this file from the website called thingiverse.com. Explanations are given for several elements to explain how nuclear shapes affect properties of elements. Thorium is explained in Thingiverse to relate its fission capability due to its salient protons.

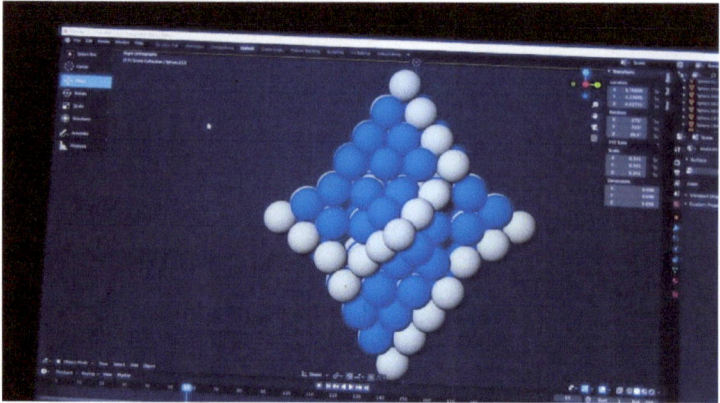

Figure 31: neodymium using a Blender file for a 3D printer. No axis proton is seen in the two rings. White protons, blue neutrons for neodymium 3D model software

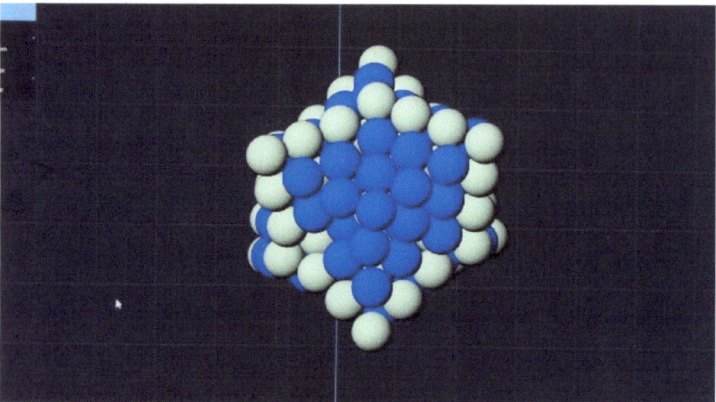

Figure 32: neodymium has two coaxial rings of protons, but no proton on the axis

Neodymium has 18 protons on each ring of white protons, in Fig. 32. When Nd is near an iron atom, the flux vortex from Fe gathers all 36 flux lines from Nd to enhance the magnetic flux density B. The 12 Fe lines of flux plus the 36 Nd lines of flux result in a 4X increase in the B field, ideally.

Once the neodymium model is finished, gadolinium can be assembled by adding neutrons into the lowest pit and then adding four protons.

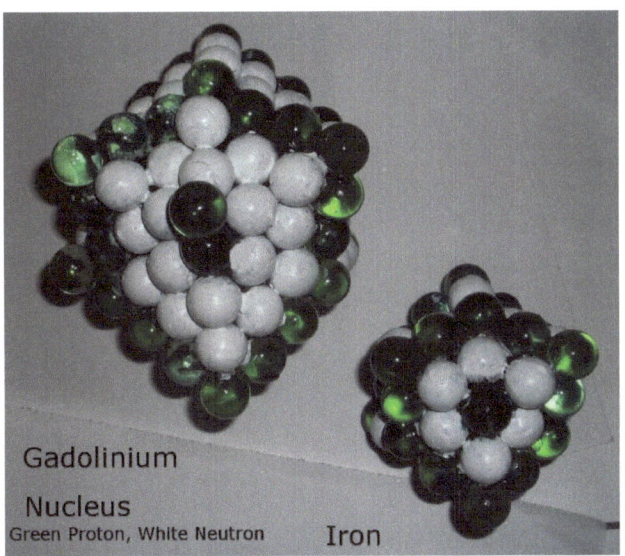

Figure 33: gadolinium and iron have similar shapes

The gadolinium nucleus has two protons on the axis of each ring (Fig. 33). That extended shape fits well into the extended ring shape that occurs due to the top and bottom faces of the cube being covered by a pyramid-4, while the four side faces have a pyramid-3. That allows a tolerable similarity to Fe. The elements adjacent to Gd in the periodic table are europium and terbium. Those two paramagnetic elements do not qualify as ferromagnetic elements above the freezing temperature because the axis protons are not symmetrical, as much as Gd is symmetrical. The tolerances for similarity to Fe allow Gd to ferromagnetic above 0 Celsius, but Eu and Tb are not. This is a judgement call on the part of the author. At cryogenic temperatures, heavy paramagnetic elements become ferromagnetic. Light paramagnetic elements, like oxygen, can never become ferromagnetic because they do not have rings of protons.

Chromium from Argon

The element called chromium is used in a new field called spintronics [4B]. This is actively being researched in Japan to make nanometer thin films produce unusual magnetic effects. The spin of electrons is credited for giving the name to spintronics, but now, chromium is understood in more detail. The nucleus of Cr has a shape that produces the anti-ferromagnetism. Two rings of 10 protons each partially overlap, as described in the following scholarly paper.

Antiferromagnetic nucleus of Cr

Alan C. Folmsbee, Haiku Road, Haiku, Hawaii 96708 USA
June 3, 2019 for the Journal of Nuclear Physics July, 2021 web-Issue

Abstract
   The protons and neutrons in the nuclei of the elements are, in this theory, positioned in fixed locations for all common isotopes. Sphere stacking of baryons is used with both cubic and hexagonal packing. It is asserted that all elements heavier than boron have a cube of nucleons in the center of each nucleus. To provide evidence for this theory, the properties of iron and chromium are compared using their nuclear structures. Cr has two rings of protons. Geometric reasoning shows it may be possible to make a Boolean logic gate out of four atoms because of the magnetic phenomena of chromium.

Summary
   This is a static nucleus theory in which chromium has ten protons in a ring and ten more in a second ring. The antiferromagnetic property is implied by the non-coaxial positions of the rings. That is compared to the rings of protons in iron and to all elements that were articulated using the same geometric Rules as were used when the Cr mock-up was assembled. It is asserted that this static theory with a simple cubic lattice is the correct theory of the nuclear structures of iron and chromium. The choice of Cr and Fe was made to relate the properties of these elements to the A and Z, the mass number and atomic number, for those elements.
Chromium is the only element that is antiferromagnetic. Iron is one of four ferromagnetic elements. That is why these two elements are

featured in this article. Their nuclei have shapes that produce those properties. Other elements do not have the nuclear structures that cause ferromagnetism or anti-ferromagnetism.

In this theory, the nucleus of chromium has structures that produce anti-ferromagnetism. This static nucleus theory uses a stacking of spherical protons and neutrons to produce a type of lattice that is called face armored cubic. A simple cube of baryons (protons and neutrons) is at the center of all elements heavier than boron. Each face of the cube is covered by a pile of baryons. That shape acts to armor the face of the cube so the cube will not be broken during a collision. That armor is usually shaped like a pyramid. The Cr nucleus is based on the Ar nucleus. Fig. 34 shows the components of argon's nucleus: a cube and six pyramids to cover the six faces. The four sides have a one-layer pyramid and the two other faces have a three-layer pyramid. The white spheres are protons and the dark spheres represent the neutrons.

Protons tend to form lines of protons in the nucleus. This is seen even in a random collection of 57 baryons for iron. The 31 neutrons and 26 protons commonly have protons touching protons. It is not possible to place neutrons to isolate protons in that model. Details were given in the chapter of "iron versus copper" and in the source of that chapter, The Journal of Nuclear Physics, March 2019 [1B]. The proton lines sometimes form rings of ten to twenty protons around the nucleus. Chromium has two rings of 10 protons in the nucleus. The 12 proton rings in Fe are compared to the proton rings in Cr using geometric reasoning. Iron's rings are coaxial but each chromium ring has an axis which passes outside of the other ring. This reveals the cause of anti-ferromagnetism and ferromagnetism for nuclei. To provide certainty, all elements are compared to Cr in a periodic table of nuclear structure.

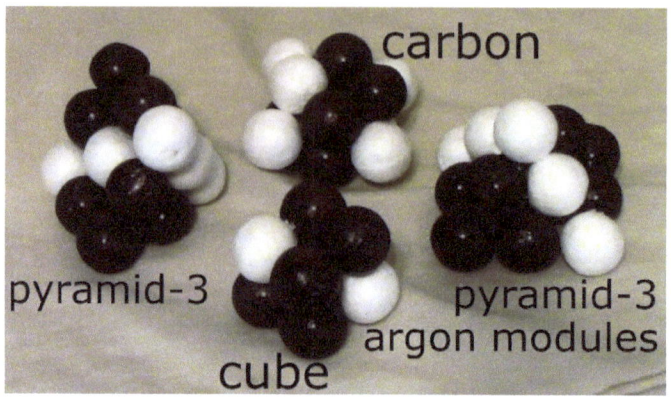

Figure 34: Argon is made from carbon and two pyramids with 3 layers, white protons

Introduction

Element 24 is called chromium and it has already been tested by other researchers. Some unusual properties have been reported for Cr. It is able to self-passivate its surface. A five-fold bond in dichromium was well publicized in *Science* [2B]. Ten electrons are said to be involved in a Cr atom using that five-fold bond [2B]. The paramagnetic property is sometimes observed in chromium. Anti-ferromagnetism is reported for Cr. The magnetization of one atom is cancelled by a reversed direction of magnetization in the adjacent Cr atom. Large chromium blocks are not ferromagnetic, but at a small scale, an atom can produce a magnetic flux that goes to a nearby Cr atom that points the flux the other way. A line of flux is defined as a line from one proton to one remote electron. Ruby lasers use Cr. Cryogenic temperatures affect an antiferromagnetic element in a way comparable to their affects in superconductors, laser devices and ferromagnets. The giant magnetoresistance effect [4B] uses a structure like in Fig. 44, with an iron layer on a chromium thin film. This paper on the nuclear structure of Cr asserts that the static allocation of baryons in the nuclei accounts for many of these phenomena. In contrast, iron is also shown with details next to chromium, with models of nuclei that are ten trillion times larger than is natural. The ferromagnetic iron is compared geometrically with the antiferromagnetic geometry of the chromium nucleus so that the reader will be convinced that this is the correct representation of reality. The static nucleus theory fits the properties of these elements well. Other elements also fit with this pyramidal cube theory well, such as technetium and uranium.

Nuclear structure has been a mystery to past researchers. Strong magnetic affects are ascribed to electrons, not to nuclei. The standard idea was that nuclei have very weak magnetic moments and the iron nucleus was negligible compared to its electrons. It was said that spins of electrons cause ferromagnetism, with little discussion of the structure of the protons being in loops. An exception is from Andre'-Marie Ampere in 1819. He said he expected those loop currents in Fe. In recent years, the structure of the Ne nucleus was calculated using Schrödinger's equation and a computer in France [5B]. The shape of Ra was experimentally measured as having a pear shape [6B]. The periodic table in this paper shows the same shapes in the theory as shown in images on those papers on neon, barium [7B] and radium [3A]. That periodic table is in Fig. 23. This pyramidal cube theory shows why uranium decays into two products with a bimodal distribution of masses. The 27-nucleon cube goes to one fission fragment or the other fragment. That is why the two mass modes differ by 27 in mass number A. There are many such instances of evidence that support the pyramidal cube theory of the structure of the nucleus.

Nuclear structure was sometimes described in books as dynamic or chaotic, without a static placement of protons or neutrons. In [6B], those authors wrote: "For certain combinations of protons and neutrons there is also the theoretical expectation that the shape of nuclei can assume octupole deformation, corresponding to reflection asymmetry or a "pear-shape" in the intrinsic frame, either dynamically (octupole vibrations) or statically (permanent octupole deformation). " Gaffney et al. In that quote, a static nucleus theory was described.

The two rings of twelve protons in Fe were discovered in May, 2017 by the author and are reported in [1B]. The proposed theory of Cr and Fe uses a static nucleus theory. A fixed stacking of neutrons and protons (baryons) is proposed in this paper. No quarks are needed to describe any kind of magnetism. Proton spheres are expected to be more than 0.83fm in radius, according to standard physics. Sphere stacking has provided realistic results in the effort to describe every nucleus.

Figure 35: the argon nucleus is the foundation element supporting chromium

The simple cubic lattice of baryons at the center of the nucleus
All elements beyond boron have a simple cubic lattice of baryons at their cores. A cube-2 is the name given to an eight-baryon stack. This cube is used inside carbon through manganese, the cubes all have two protons and six neutrons, as shown in Fig. 26. Protons are white and neutrons are dark. Four modules are shown for sphere stacking: a cube. A carbon nucleus, and two pyramid-3 modules for argon. The cube has two layers of protons and neutrons, so the geometric module is named cube-2. That cube has two protons and six neutrons, evenly distributed so the two protons are far from each other. The carbon nucleus is a cube-2 with four protons added onto four of the six faces. The protons tend to make lines of protons as they are added one at a time to assemble Cr. First, the Ar nucleus is made as a foundation element for Cr to be produced by adding six protons and six neutrons.

   carbon = cube + 4 protons
   Ar = carbon + 2 pyramids
   Cr = Ar + 6 protons + 6 neutrons

Those simple equations express how the mock-up was produced. Here is a more detailed version of how the chromium mock-up was assembled.

Carbon.12 = cube + 4 protons

Ar.40 = carbon + 2 pyramids

K.40 = Ar changing a neutron to proton

Ca.40 = K changing a neutron to proton

Sc.45 = Ca + 4 neutrons in deepest pits, 1 proton in line with protons

Ti.48 = Sc + 2 neutrons in deepest pits, 1 proton in line with protons

V.51 = Ti + 2 neutrons in deepest pits, 1 proton in line with protons

Cr.52 = V + 1 proton in line with protons

The added neutrons are allocated to Ar due to isotopic facts and by estimating hydrodynamic sinks in the structure of the nucleus. Deep gaps are allocated with neutrons as if gravity pulled all nucleons together. Each nucleon is treated as a sink for a liquid and nucleons fill the gaps in the stacked face armored cubic (FAC) lattice. Symmetry is maintained during allocations. Proton allocations are additionally constrained by the trend for protons to join lines of protons. As each element, from Ar with Z=18 to Cr with Z=24, is fabricated, the mock-up benefitted from the fact that the first three elements all use isotope with A=40. In the periodic table in Fig. 23, the reader can see that Ar, K, and Ca all use isotope 40, so they all have the same silhouettes.

The argon nucleus is shown in Fig. 35. This is a foundation for chromium, upon which a succession of elements will be made. The potassium mock-up has one neutron changed to a proton so A, the mass number, is 40 for potassium. Argon is seen to have carbon in the middle, with its simple cubic center. On the top and bottom faces of C, two pyramid-3 modules are placed. Each pyramid has six protons and eight neutrons. The cube in Fig. 36 is seen with the <011> crystallographic plane displayed. The pyramids are nestled onto the cube, since 3 layers fit onto 2 layers snugly.

Figure 36: the six faces of each cube have armor to be stable. Fe is a pyramidal cube. So is Cr.

The completed chromium model is shown in Fig. 36, next to the iron model. The three axes are marked in yellow highlights. The top axis on Cr comes out of a ring of 10 protons. The lower axis on Cr emerges from a place outside of the top ring. It comes from the center of a hidden ring on the rear side of the Cr model. The iron model, on the left, has one axis emerging out of the center of a ring of 12 protons. That axis is also passing through a second ring of protons that is on the other side, out of sight. The two rings in Cr have a mutual area that is less than half of the area of a ring. If flux is emitted evenly from each loop, then there is simultaneously a mutual flux and two external flux bundles. The author considers that electrons are paired with each of these protons, but that bond is not shown in this paper. Some remote electrons are paired with a subset of the Cr protons. Some atomic electrons are paired with the local protons. Some local electrons in one atom have a bond that passes through a local ring and which is terminated on a remote proton. In this limited way, it is like iron as a ferromagnet. Iron has a flux from one ring passing through the second ring locally. Chromium has a flux from one ring passing through the second ring remotely. That defines anti-ferromagnetism at the particle level.

Five face views of the cube inside a Cr nucleus are shown in Fig. 37 and Fig. 38. Alongside Cr is the Fe nucleus so the reader can see how a ferromagnetic integer topology differs from an antiferromagnetic integer topology. Iron uses a cube-3 core with 27 baryons [1B] while chromium uses a cube-2 with 8 baryons. (See the chapter on "iron versus copper"). Fig. 37 has Fe on the left and Cr on the right as the four faces are shown together, labeled front, left, rear, and right. The axis through Fe is labeled f and the two axes through Cr are labeled r and c. The Cr model has red numbers on the six protons that were placed onto the argon foundation nucleus. The number 19 is on a proton to make K and the numbers 20 through 24 are on the proton that was added to Ar to get, respectively, Ca, Sc, Ti, V, and Cr.

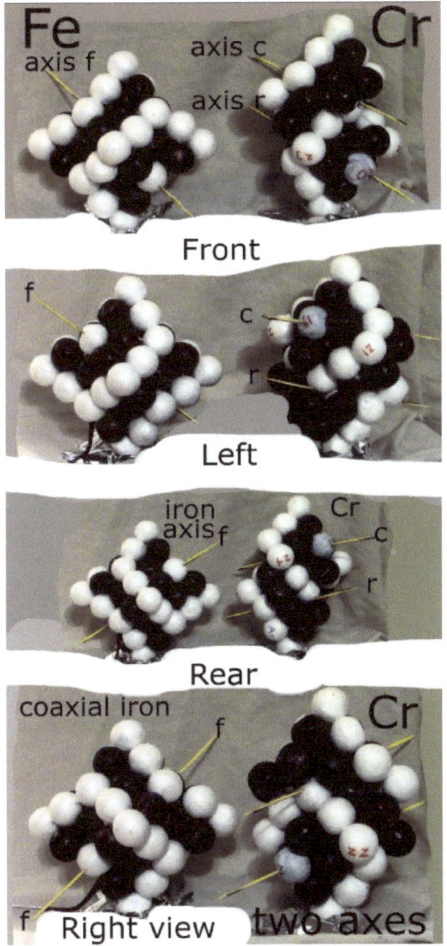

Figure 37: ferromagnet versus antiferromagnet: four views

Six faces of the cube are armored with baryons

The chromium nucleus has a cube with 2x2x2 baryons. That cube-2 then becomes carbon by adding four protons on four faces of the cube. Carbon then becomes argon by adding armor to the remaining two faces. A pyramid of baryons with three layers is nestled onto the two-layer cube's top face and another is on the bottom face. Protons tend to form lines of protons.

A hexagonal close pack lattice was evaluated as an alternative to the pyramidal cube theory of nuclear structure. This hexagonal lattice has disadvantages because it is finite. The lattice is terminated and a then the hexagonal close pack produces a porous exterior. The gaps between the nucleons are as large as on a cubic lattice. The armoring on a cube produces small gaps as pyramids are stacked. In summary, a cubic core gets an outer surface of a hex lattice but a hexagonal core gets a cubic exterior. That allows the cubic core to survive and the hex core to be destroyed during nuclear collisions.

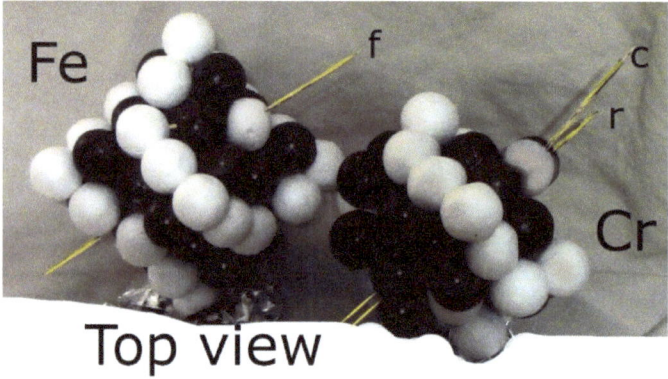

Figure 38: protons form lines of protons

Argon is the foundation element for chromium

There are eighteen foundation elements. They are defined as elements with:
1. a cube with pyramids of baryons on all six faces
2. with all four sides being the same
3. pyramids have up to six layers or zero for carbon
4. 90 incremental elements are based on 19 foundation elements

The foundation elements are: C, O, Ne, P, Ar, Fe, Ge, Kr, Zr, Cd, Xe, Ce, Hf, W, Po, Rn, U, Md, Nh. Notice that all noble gases are foundation elements. Argon is the foundation element for chromium. Six protons and six neutrons get added to Ar-40 to build the incremental element Cr-52. All elements are incremental elements, except for the foundation elements and elements lighter than carbon.

Here are four the Rules that were used for constructing an argon mock-up for later building chromium:
A cube of baryons is designed at the center of the nucleus.
Protons in the cube are far from each other.
Outside of the cube, protons tend to form lines of protons, seeded from the cube.
A pyramid grows to completion on each face of the cube.

Incremental elements building up for chromium
The Argon foundation has 40 baryons with 18 protons and 22 neutrons. Potassium and calcium also have A=40. That is a clue that those three elements should use, in this evaluation, A=40. As a result, two neutrons were converted to protons. In the periodic table of nuclear structure, elements 18, 19 and 20 have the same silhouette. The element with 21 protons is scandium and its isotope has 4 more neutrons and 1 more proton than calcium-40. The mock-up was examined and the low points in the Argon pyramidal cube were treated as hydrodynamic sinks. The nucleons were place to fill the holes and to continue lines of protons. Fig. 39 shows how the lines of protons become two rings of protons for Cr. These rules are made because hydrodynamics is expected to be the most realistic phenomenon responsible for the structures of nuclei. Notes about the rules are in [1B] with images of mock-ups of Cu, Ni, Co, Gd, Ne, U, Hg, Zr, and Tc. Also, the book by this author is useful. It is entitled, Charge distributions on the nuclei", ISBN 979 836349 5403. That book has an explanation to justify each of the 19 Rules.

Figure 39: iron is like a coaxial connector shape, Cr is like a parallel port

The two rings of protons in Cr

A sketch is provided in Fig. 40 to give a schematic simplification of the Cr nucleus, as it relates to magnetic phenomena. No neutrons are shown. Ten protons make one ring and ten more make the second ring in Cr. That number is related to the five-fold bond by expecting five pairs of electrons to be involved in the chemical bond. The two rings of iron are sketched (Fig. 41) to compare to the two rings of chromium. The rings have an undulating shape because of the pyramidal cube structure of nuclei heavier than boron. The six faces of the cube-2 are armored by symmetrical piles of nucleons. The face armoring is also called a pyramid. The argon foundation provides two lines of protons on two faces of the cube in Fig. 35.

Predictions from the Static Nucleus Theory

This article will show that each loop of ten protons in chromium can receive independent magnetic flux bundles from remote locations outside of that particular atom. Each ring can receive two parallel fluxes with opposite spins. If the spins are polarized the same way, the two flux vortexes drive the mutual area of Cr to emit flux. If the two spins are not going the same way, no flux is emitted. This can make a chain of magnetically entangled Cr atoms, without a macroscopic magnetic flux. The two input ports from that situation can allow a possible Boolean logic function to be done. The result from a Cr nucleus is a new flux from the place where a mutual area is made between the two rings. This is a special place because torque is available. One proton ring is a fulcrum and one ring is the source of magnetic flux. The electron is at the other end of this flux.

Iron has twelve protons in each loop. The direction of the ring currents can be set by electron eddy currents. At high temperatures, the electrons disrupt the nuclear ring currents so the iron is demagnetized. At low temperatures, demagnetization is done by having electron eddy currents drive the proton ring currents in two opposite directions. Since these rings are coaxial, the external flux that sets the two ring currents must enter from opposite directions to set the currents to oppose each other. In contrast, chromium can accept two parallel fluxes to the two loops. That is the difference between ferromagnetism and anti-ferromagnetism. Iron is less influenced by parallel opposite fluxes than Cr. Iron will adapt to those parallel inputs by having both ring currents go the same way. Cr will adapt to parallel opposing fluxes by easily having the two ring currents go the opposite ways. At high temperatures, Fe and Cr have disrupted ring currents due to electron temperatures influencing the nucleus.

Significance of this allocation of protons in chromium

The theory of the static nuclear structure is given persuasive evidence for its correctness by the facts about chromium, iron, and gadolinium. The Cr nucleus has two axes for two rings of protons. The first axis is tangential to the second ring. This can be interpreted as the axis being important for flux. An alternative way of using this geometry is to give importance to ten discrete flux lines from ten protons to ten electrons. In the second case, there are at least two lines of flux that are mutually inside both loops. That is like a ferromagnetic part of Cr. The other protons in the loops can be considered to cause anti-ferromagnetism by cooperating with other chromium atoms that are nearby. Then, a mutual flux can be made, or a reverse flux can be made. Figure 44 shows one situation where Cr atoms are chained by flux that needs two input fluxes with the same spin to have a mutual flux emerging from the middle of the nucleus. This understanding uses the shape of the nucleus in the schematic. The iron nucleus is also shown in profile to feeding magnetic flux to a chromium nucleus by way of electrons in eddy currents.

Fig. 42 shows a simplified side view of the chromium nucleus compared to iron in Fig. 43. No neutrons are shown so the proton loops are clear. The blue loop is over a white loop in Cr and the mutual area of those loops is small. Iron has a large mutual area. The idea is that Cr has two antenna areas and one mutual flux area in the middle. The antenna areas with a single loop area are available to intercept flux to set the direction of ring current. If both antenna lobes of the rings are intercepting two fluxes that go opposite spins, then the mutual flux is zero. If the two inputs to the two antenna zones are spinning the same way, the mutual flux is sent outwards as a vortex of flux.

In Fig. 42, the side view of the protons in Cr are shown. Below that is iron in Fig. 43. These shapes are used in Fig. 44, the schematic of anti-ferromagnetism.

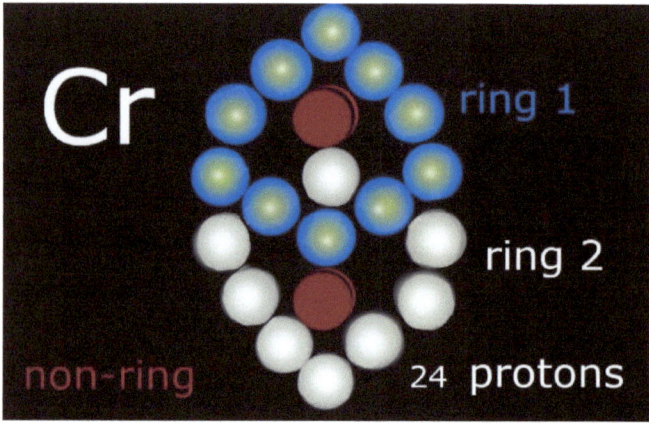

Figure 40: the two rings in chromium, ten protons in each ring, neutrons not shown

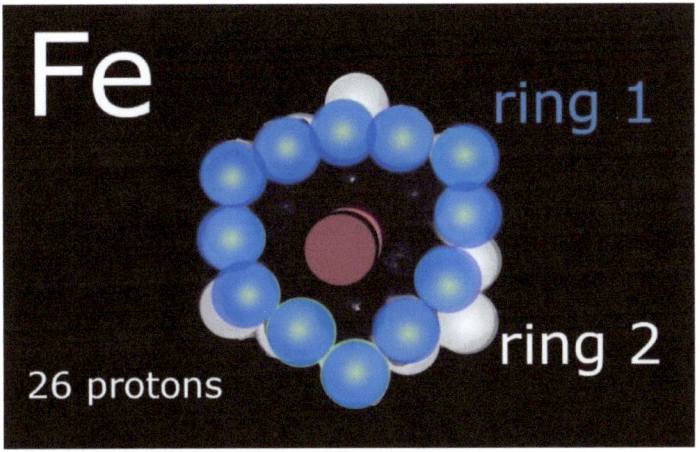

Figure 41: iron has one axis for two rings, view from <111> crystallographic axis

The schematic of anti-ferromagnetism
In Fig. 44 are some Fe atoms and Cr atoms. Little versions of Fig. 42 are used in this figure for the Cr nuclei. Also, small copies of Fig. 43 are used to show the side view of the Fe nuclei in Fig. 44. The flux from the top left Fe nucleus is sent to a Cr atom. Each red electron is paired with only one proton. The proton ring currents drive a north pole flux vortex, if the ring current spin is clockwise (CW). Black is for north poles, and all black lines are from nuclei to electrons. Red lines are for south poles where electrons drive a nuclear ring current spin direction. The lines do not cross, they are entwined so a rotating bundle of lines drives a different bundle of lines from a different electron eddy current, which is paired with a remote nuclear proton ring. This is the non-crossing law that the author proposes to be consistent with Lenz's Law. Nuclear ring currents going CW are paired with electrons that are moving CW in an eddy current.

The path will be traced from the top left Fe going down to Cr and bending right to go to the top right corner of Fig. 44. In that way, flux is reversed. That is anti-ferromagnetism. Random orientations of many Cr atoms will statistically negate any macroscopic directionality of flux.

The Fe atom at top left in Fig. 44 sends flux to the left and right antenna of the Cr nucleus. That drives currents in both rings of Cr, so the mutual area emits a flux. That flux goes to another Cr atom, and that is chained onwards to other atoms. The two input ports at the bottom of the schematic are available for quantum computation, Boolean logic, frustrated magnets, rapid reversal of magnetization, and self-passivation. A Boolean function might be made from four atoms in Fig. 44. A two input NAND gate could be made, although, without amplification at each stage, the logic signal would dissipate. More than four atoms, for example four billion atoms, might be used for one NAND gate. This would use thin layers of Fe and Cr to separate the logic inputs from each other. Test devices could be built with connections made with macroscopic sizes, until smaller connections become achievable.

Notice that the CW flux from the Fe nucleus is sent to make an eddy current of electrons that is also CW. The north magnetic pole is from the protons to electrons and it is shown in black. This is using a CW flux vortex. That north pole is bent on a path on Fig. 44 so it makes a 180 degree turn through several Cr atoms. The flux emerges at the right top side of the figure as a north pole (N) with counter clockwise (CCW) property of spin. The south poles are in red and they are from an electron eddy current to a proton ring current. Each eddy current is driven by a nearby second eddy current.

The result of Fig. 44 is seen where the north magnetic pole has the N letter with an arrow pointing down on the left, and on the right side of the figure, another N is there with an up arrow. That means a north pole was going down, but another north pole is less than a nanometer away and it points up. The spin is CW on the left and on the right the spin is CCW. That change in polarity is cause by random orientations of Cr atoms.

Comments on chemistry of Cr versus Fe in the s orbital

The education in chemistry uses the 1s orbital in all elements. The new understanding of the nucleus shows that to be justified for elements up to Mn, with $Z=25$. The cube-2 in those light elements has two protons. That is why a 1s orbital is accurate to hold up to two electrons. Iron and some middle weight elements have a cube-3 with 8 protons, not 2. Therefore, the author expects that iron through bromine may be modeled without the 1s orbital being limited to two electrons. A new 8 electron s shell is recommended. This illustrates how the static nucleus theory can be related to several phenomena.

Elements starting at Fe do not have 2 protons in the core, they have 8. This implies that chemistry can change its description of the transition elements so the 1s orbital is obsoleted. Elements from manganese downward can use 1s orbitals to fit the two protons in the cube. Elements from iron to iodine should have no 1s orbital, but should use eight electrons in the lowest shell to match the eight protons in the central part of the nucleus.

The periodic table of nuclear structure is in Fig. 23. The silhouettes of 118 elements are shown using color coded baryons. The red circles are baryons in a simple cubic lattice. The black circles are baryons. Blue circles are added onto incremental elements so that gaps in the foundations are filled. A prioritization of placements of the blue circles is intended to reflect a hydrodynamic flow into each baryon. The shapes of nuclei have also been evaluated in simulations [5B] and experiments [6B],[7B]. Please notice that the two radioactive elements Tc and Pm have a cube-4 structure. That is why they decay more readily than elements adjacent to them on the periodic table.

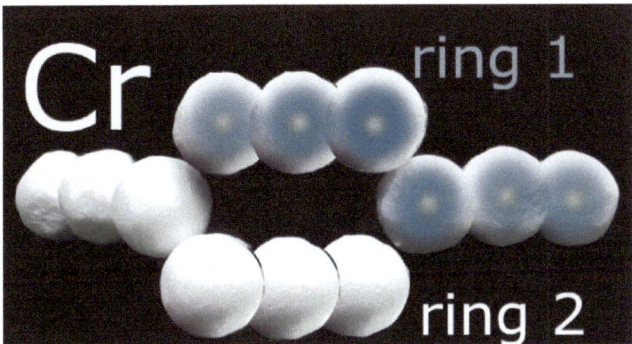

Figure 42: protons in chromium rings, side view with no neutrons shown

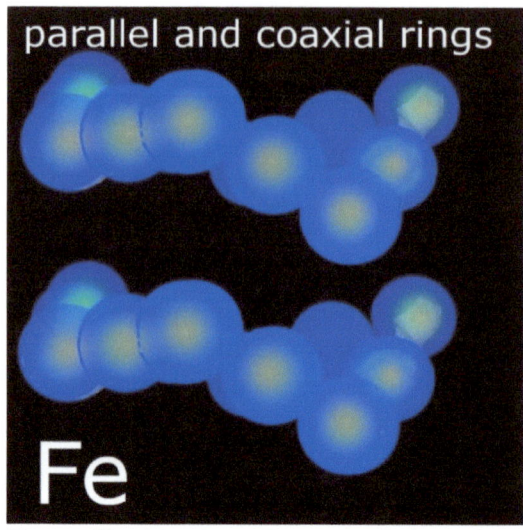

Figure 43: side view of protons in iron rings, no neutrons shown

Anti-ferromagnetism and the rings of chromium

The only element that is antiferromagnetic is Cr. That category is defined by its ability to have local magnetic flux near atoms, but to then negate that flux with a negative flux. This is called frustrated magnetism by some authors [4B]. The nucleus of chromium, the author asserts, is a static lattice called face armored cubic. The two loops of ten protons provide a plausible explanation for the unique antiferromagnetic property which no other element possesses. The north pole was defined as being sourced from protons in Cr nuclear rings that carry a clockwise current. Those protons are paired with remote conduction electrons which are also the north pole. A line can be drawn between the paired particles, like in a bond path in chemistry [8B]. That reference describes that line between two atoms as an exchange channel. In the current paper, this is made to specify also that one proton in one atom has a bond path to an electron, and that this is where an exchange process is involved in anti-ferromagnetism. Figure 44 shows the black lines as north and the red lines as south magnetic poles. That means one proton is north and one electron is north because the spin of the nuclear ring current is clockwise. That nuclear current in 10 protons is tracked by each of 10 electrons. The 10 lines form a circling vortex of lines from a Cr nucleus to an eddy current. The motion of the north electrons drives nearby electrons of the south pole. Those electrons connect the south pole to protons. Those protons also have a clockwise ring current, relative to the electrons. Conversely, that south pole proton ring current has a counterclockwise rotation relative to the nucleus.

The positioning of protons and neutrons used hydrodynamic expectations and goals of symmetry for stability. Integer geometries were based on A and Z, the mass number and the atomic number. This gives a geometric reason for spin polarization. The rings of protons have a ring current that makes a magnetic flux. This flux vortex is made lines from protons to electrons. For chromium, some bond paths make a mutual flux in both rings and some parts of each ring do not emit a flux vortex. Those protons are paired with electrons without using a flux vortex. There is no torque available from single ring of protons, compared to the two-ring topology.

The two rings give torque to some remote electrons. That structure provides a chaining into nearby chromium nuclei with opportunities for unique interactions. Self-passivation may be enhanced by this flux chaining from one Cr atom to the adjacent Cr atom. A distinction is made between a paired proton-electron and its bonding theory as compared to a flux vortex which is always from a loop current with a torque enabling topology. The pair theory is for every proton. The flux theory is only for the few pairs that have their fluxes originated by nuclear ring currents or a macroscopic loop current. Large inductors have torque without needing iron ring currents. The electron current alone can make torque because billions of atoms are supporting each other as fulcrums. In iron and chromium, the two ring currents are set by external electron conditions. Once started, the currents persist without external drive. Chromium is more controllable than iron for changing the magnetization polarity because it is susceptible to parallel fluxes of opposite polarities. Iron's susceptibility to opposing fluxes only comes it the two fluxes are originating from opposite directions. The pyramidal cube theory of the nuclear structure of chromium provides a fertile area from which many applications can be produced.

Conclusion for Cr

This paper has two goals. The first goal is to use a unique element, chromium, to add to the confirmation of the pyramidal cube theory of the structure of the nucleus of all elements. The second goal is to announce that anti-ferromagnetism is caused by a topology in the nucleus of Cr. The definition of a north pole was given relative to a nucleus and its associated electrons. A new definition of a line of flux was given. The torque from a magnet was explained as using a fulcrum in iron or chromium that is based on the two rings of protons. That is why magnetic flux is different from electric attraction. The spinning of the nuclear ring current keeps the electrons spinning with the same polarities. All of these insights became clear after confidence was achieved in the correctness of the static nucleus theory.

Acknowledgements for Cr

The author wishes to thank Professor Mahmoud Melehy and professor Faquir Jain of the University of Connecticut for revealing the state of the art of physical electronics up to 1989.

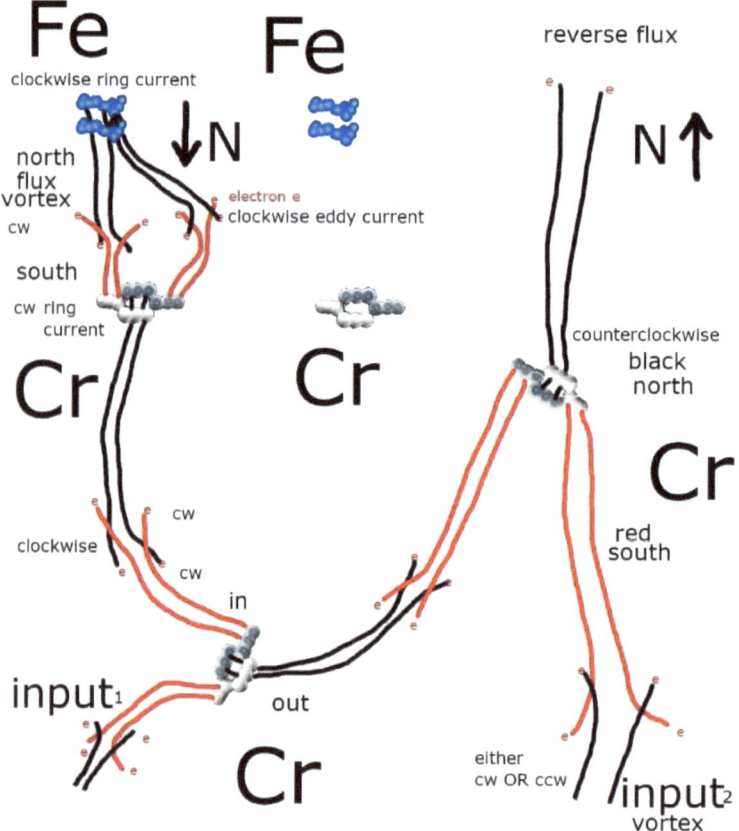

Figure 44: Schematic of antiferromagnetic functions of each nucleus

References Section B

1B A. Folmsbee, *Journal of Nuclear Physics*, March, 2019, Magnetism from iron's nuclear structure

2B T. Nguyen, A.D. Sutton, M. Brynda, J. C. Fettinger, G. J. Long and P.P. Power. Synthesis of a Stable Compound with Five-Fold Bonding Between Two Chromium (I) Centers, *Science*, 2005, 310, 844-847

3B G. Frenking, S. Shaik. The Chemical Bond, volume 2, pp 259, 2014 Wiley

4B K. Sato, E. Saitoh, Spintronics for Next Generation Innovative Devices, pp 1 and 82, 2015 Wiley-VCH

5B J. P. Ebran, Elias Khan, T. Niksi and D. Vretenar, *Nature*, July 19, 2012, How atomic nuclei cluster

6B Gaffney, Butler et al. May, 2013, *Nature*, Studies of nuclear pear-shapes using accelerated radioactive beams

7B B. Bucher, *Phys. Rev. Lett*. 116, March, 2016, Direct Evidence of Octupole Deformation in Neutron-Rich 144 Barium

8B G. Frenking, S. Shaik. The Chemical Bond, volume 1, pp 291, 2014 Wiley

Promethium and technetium

The element promethium has a shape that was not planned by the author to be promethium. The shape was planned to be analogous to iron, but using more layers of baryons. Since iron was convincingly correct in its nuclear structure with a cube-3 and six pyramid-2 modules, it was reasonable to use inductive logic to extend the plans for one more increment. A cube-4 (Fig. 45) with six pyramid-3 modules nestled onto the faces of the cube was attempted. By adding up the 64 baryons in the cube and the 14 baryons in each pyramid, the total was 148 as the mass number A.

$$A = 64 + 6*14 = 148$$

When the 148 isotope was considered, it was not known which element it was appropriate to use. The elements neodymium, promethium, and samarium all have an isotope 148. Prometium-148 has a half-life of 41 days. Pm-147 has a half-life of 2.6 days. The author considered that the cube-4 could have an instability that would recommend Pm over the Nd and Sm isotopes. The next consideration would be about Z, the atomic number for promethium [10C].

To calculate a Z of 61 protons, first a simple calculation was done using the simplest cube-4 plan and the simplest pyramid-3 plan. The 4x4x4 cube has 64 baryons. If half of them are protons, then there would be 32. The pyramid-3 can have 1 + 2 + 3 protons on its three layers. Adding all of those gives:

$$Z = 32+6*6 = 68$$

The resulting 68 protons is 7 more than the Z=61 for promethium. It was decided to eliminate 7 protons from the modules of pyramids and a cube (Fig. 46 to Fig. 51). The cube can have 4 protons removed from its 8 corners, for symmetry reasons. Then 3 protons can be taken from three pyramids out of six pyramids, for symmetry reasons. That gives the 61 protons for Pm. A model was built out of Styrofoam spheres to determine if the shape looked reasonable. The next figures show those models.

Figure 45: promethium modules by iron modules

Figure 46: Iron is the prototype for promethium cube-4 and pyramid-3

Figure 47: one pyramid added on the analog of iron, promethium modules assembly

Figure 48: three pyramids added on cube-4, nestled into pits between spheres

Notice that the two pyramids at the bottom of Fig. 47 are different from each other. One has three white protons on its base and one pyramid has two white protons on its base. That was done to get Z=61 while maintaining symmetry. Three of the pyramids have 5 protons, each, and the three remaining pyramids have six protons in each pyramid. That allows symmetry to be used. The cube also was given a symmetrical change to get the number of protons right for Pm-148. The cube has eight corners. Four of the cube's corners would have a proton if a checkerboard pattern of proton allocation were used. By removing those four protons, Z is 61, as needed. Also, that results in all eight corners of Pm being occupied by neutrons. See Fig. 46 or Fig. 45 to verify that brown neutrons are in every corner of the cube.

Figure 49: loop of white protons in promethium

The author deliberately put symmetry into the nuclear structure of these sculptures to mimic how nature favors symmetry in the production of isotopes. The final models in Figures 26, 27, and 28 show that this planning has given the Pm-148 a simple shape.

Figure 50: promethium has a 3-way proton intersection roundabout

Figure 51: promethium-148, oriented as silhouette, like in the new periodic table

After Pm looked good, Tc (Fig. 21) was also modeled with a cube-4 so the instability of that module would be applied to several unstable elements. Those elements turned out to be oganesson-294, protactinium-230, promethium-148 and technetium-99. No other elements use the cube-4.

Platinum and vanadium catalysts

The most famous catalyst is platinum. But several other elements are used by chemists to promote chemical reactions without consuming the catalysts very much. Gasoline powered cars use platinum and palladium in the exhaust pipes to burn off any unburned fuel. Those catalytic converters are so valuable that they are stolen by thieves, commonly. Platinum has a price of more than $1100 per troy ounce. Pictures of a platinum nuclear mock-up are in Fig. 54 and Fig. 55. Pt costs about $14,000 per pound. Vanadium (Fig. 56 and Fig. 57) is a catalyst for some chemical reactions and it costs about $14 per pound.

The heavy metals like Pt and Au are based on the foundation element, tungsten. All 10 elements in the tungsten class have a 6-layer pyramid on the top face of the cube-3. That provides armor for the cubic lattice by having the hexagonal close-pack lattice (HCP) on the surface. The gaps between baryons in the HCP armor are about four times smaller than the gaps in the cube. That prevents incoming particles from splitting the nucleus, usually. Fig. 52 shows the HCP lattice on the top view of W, tungsten.

Figure 52: Tungsten top view, foundation element for platinum

Figure 53: Gold, rear view

More information about the W class of elements is in the hardcover book, <u>Charge distributions on the nuclei</u>.

Figure 54: platinum front view, with a six-layer pyramid facing the viewer on top

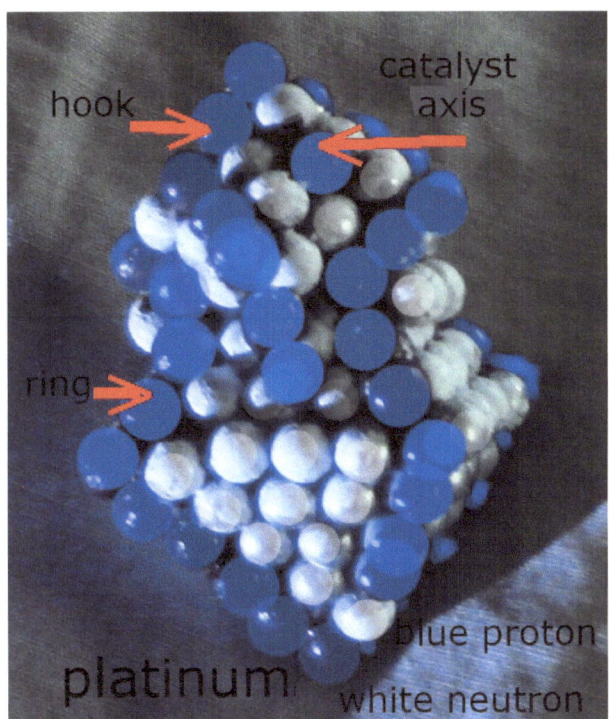
Figure 55: platinum nucleus with hook, ring and axis protons in blue, white neutrons

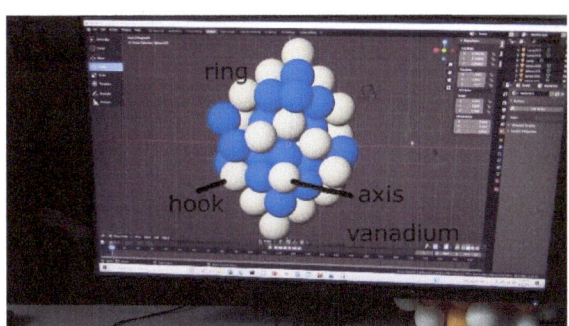
Figure 56: Blender 3D printer file of vanadium catalyst hook, and axis in front, ring is hidden in back

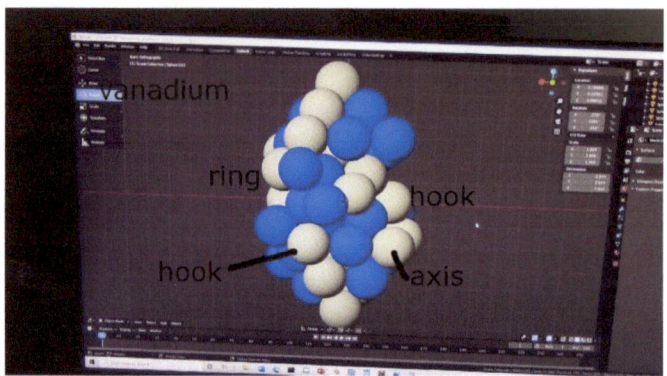
Figure 57: vanadium 3D model file for catalyst proton ring in white, blue neutrons, hook and axis protons

Evidence of the correctness of this theory

There is abundant evidence that the static nucleus theory is the correct theory of the shape of matter. A correct theory gives correct-seeming answers quickly and easily to old questions and new. An incorrect theory gives correct answers rarely. The ease with which correct-seeming answers are obtained is one indication that this pyramidal cube theory gives the correct description of nuclear structure. Experimental evidence is difficult to get for nuclear structure. Even a vague silhouette of a nucleus requires a microscope with a magnification of at least fifty billion. An atomic force microscope only has a magnification of five million. A collider experiment in France was able to see a pear-shaped silhouette for radon, and radium [3A]. Its magnification power is about eighty billion, using gamma rays that were emitted from collisions at CERN.

The vacuum chambers at Argonne National Laboratory were used during the year 2016 for barium experiments. The pear-shape nuclei were reported in the press, including on the internet website for phys.org. Radioactive barium ions collected from fission fragments of californium nuclei (CARIBU source) were accelerated with the ATLAS accelerator. See B. Bucher et al. Direct Evidence of Octupole Deformation in Neutron-Rich, Physical Review Letters.

Liverpool University on the island of Great Britain reported on their accelerated radioactive beams of radon and radium that show the octupole deformation with a pear shape. That was published in the journal Nature on May 9, 2013. The numerous authors included L.P. Gaffney and P.A. Butler. They reported strong circumstantial evidence from the REX-ISOLDE facility at CERN (the European Council for Nuclear Research) were used to get gamma ray data. Here is a link to that copyrighted image of a blurred pear shapes radon nucleus, magnified 80 billion x.

In 2017, the author independently derived pear-shaped nuclei for those same three elements, never having seen the earlier experimental images (Ba, Ra, Rn). The pear shapes were fabricated so a pyramidal cube structure would be stable. At the heavy end of the nucleus, a six-layer pyramid was placed to armor one end of the nucleus against collisions. The three elements listed above are the only elements known to this author which had experiments that show a non-spherical shape of a nucleus. Here is an image (Fig. 58) of the theoretical shape of barium using the pyramidal cube theory:

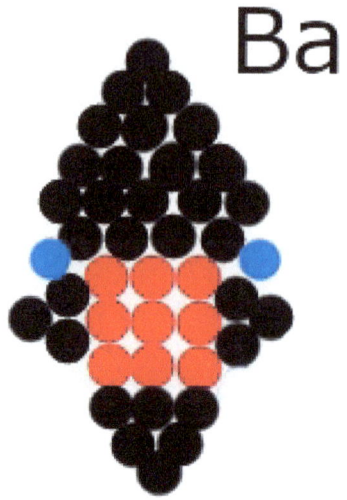

Figure 58: theoretical shape of barium having a pear-shape, magnified 8 trillion x

Barium is useful to illustrate how the experimental pear shape is related to the theoretical shape and to the function of the pear shape. In a supernova explosion, a one-sided blast will hit many candidate isotopes. The survival of barium is due to its pear shape. It is like a dart in a blast. If the pointed end of the dart faces the blast, it could survive. If the tail end of barium were facing the blast, it could be damaged. A figure is provided to emphasize this idea.

experiment, theory, and function

Figure 59: Barium has a pear shape, according to an experiment (left)

The fuzzy pear shape in Fig. 59 is not from the copyrighted paper, it was redrawn by the author. The experimental shape from Argonne National Laboratory was published in Physical Review Letters and is not shown in this book: Phys. Rev. Lett. 116, 112503 (2016) - Direct Evidence of Octupole Deformation in Neutron-Rich $^{144}\mathrm{Ba}$ (aps.org)

A synopsis article shows a picture. Heavy barium nuclei prefer a pear shape (phys.org)

https://phys.org/news/2016-06-heavy-barium-nuclei-pear.html

Neon also provides confirmation that the static nucleus theory using the pyramidal cube structure is consistent with the Schrödinger's Equation simulation in Croatia and France:

https://phys.org/news/2012-07-atomic-nucleus-fissile-liquid-molecule.html#nRlv

That paper shows the blurred probability map for a silhouette of the neon nucleus. Since that image is copyrighted, this book cannot compare their science with the author's scientific theory. Lawyers might sue scientists who want to show reasonable comparisons in public. So, the author has drawn a similar blur (Fig. 60) to show instead of the original blur from France and Croatia. That European image can be seen if you can find it using this information provided by Institut de Physique Nucléaire d'Orsay (Université Paris-Sud) and from CEA (the French Atomic Energy Commission), in collaboration with the University of Zagreb, Croatia. "How atomic nuclei cluster", J.-P. Ebran, Elias Khan, T Nikšić and D. Vretenar, *Nature*, 19 July 2012. Here is a comparison for neon silhouettes:

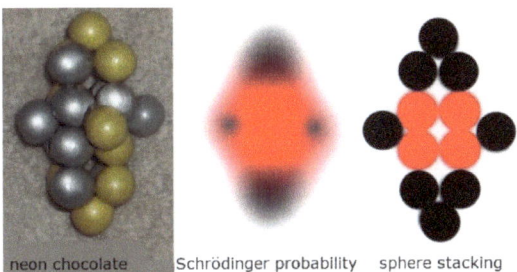

Figure 60: neon silhouettes using three methods to represent a nuclear structure

The journal Nature has the original probability distribution for the existence of baryons at certain location in the neon nucleus. Fig. 60

does not use their copyrighted image, but this figure shows a similar blur, as is common when using Schrödinger's Equations in computer simulations. Please take a look at it on this website. The atomic nucleus: fissile liquid or molecule of life? (phys.org)
   https://phys.org/news/2012-07-atomic-nucleus-fissile-liquid-molecule.html#nRlv

That website has the image that is not shown in this book.

It is obvious, when the pictures are compared, that the new theory of the pyramidal cube gives exactly the same shape as that provided by the scientists in France and Croatia. In Fig. 60, the middle image is just a blurred edit of the right image using Gimp software. The image in the journal Nature looks like that, too.

Electrons forcing protons into line

There is an equal number of protons and electrons in the universe. Pairing always exists between one proton and one electron. Each electron is paired with only one proton. Interactions between those two particle types falls into two categories: direct and induced. Each proton directly interacts with only one electron. Each pair uses induction to affect other pairs, if they are adjacent to the first pair. A line is drawn from one proton to one electron. It is a space-like dimension. A second line is also connecting each proton to its paired electron and it is a space-like dimension. A third line is drawn from the electron to the proton and it is time-like. Those three lines are combined into a contra-flow membrane called a line of flux. Two lines of flux can wrap around each other to form a chemical bond (Fig. 61).

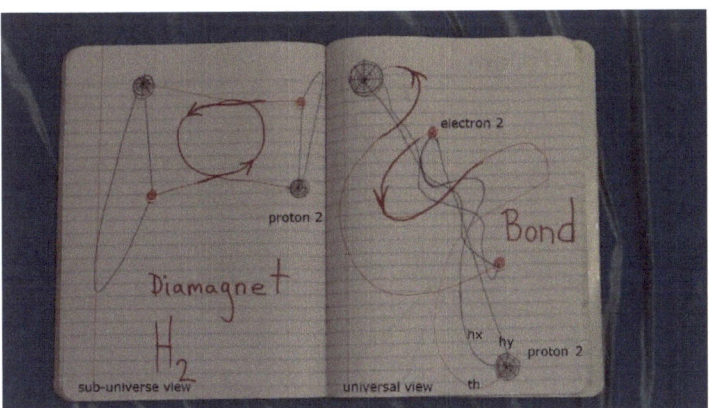

Figure 61: A chemical bond seen in two ways, sub-universal view and our normal view in the universe

The two particles can be viewed in two ways: first, as a flat sub-universe, and secondly, as a part of an atom or ion in the universe. The two views are useful for people to visualize a simple picture or a more realistic picture. Also, there are analogies that can be used to understand the two realms. For gravity, three space dimensions flow into each proton as a liquid. An equal amount of time is emitted by the proton. It is a balance of nature. Five nanoseconds are emitted while a proton consumes its own volume from the adjacent space [15C]. The analogy is then provided for the ion or atom, as in the bigger universe. Two space-like dimensions enter into the electron while one time-like dimension is sent back to the proton. Imagine an observer standing on the surface of an electron. The sky above is one proton at a distance of negative infinity. That is analogous to a person standing on the Earth looking up at infinity as a black sky at night.

There is a new type of Pauli Exclusion Principle called the non-crossing law. This states that the line of flux between an electron-proton pair cannot pass through any other line of flux without an ohmic cycle. That ohmic cycle dissipates power if it occurs. Usually, the lines do not cross through any other lines, but the pairs push each other around in a motion called "magnetic induction". A magnet has lines of flux pushing other line around to induce voltages. A voltage is increased where the density of electrons is greater than the density in another region. Also, this pairing and non-crossing law is what makes Lenz's Law. That is because when a positive charge goes one direction, a negative charge goes the same direction. In standard electrical jargon, Lenz's Law proposes that a positive current will induce a negative current. In other words, a positive current induces a voltage that opposes the first current. The non-crossing law enforces a collective phenomenon that is all about an electron and a proton being pairs using a line of flux. A line of flux has 3 dimensions: hx, hy, and th. Those are two space-like dimensions and one time-like dimension. It is like a conveyor belt. It has velocities in two opposite directions. This is true even in a star, during fusion of new elements. The electron has a flat membrane that excludes other flat membranes from crossing it. That allows electrons to force protons into a constrained motion that results in protons forming lines of protons, as flat membranes get stacked without crossing each other. Without those electrons pairing to protons, the protons would not be constrained to make lines of protons. This is a profound new insight in physics. Take notice.

It is important that these concepts are emphasized to the highest degree which the reader will tolerate. The electrons are always paired with protons, one to one. Even in the Sun, with all its violence, turbulence, pressure and its highest temperature, each proton has a line of flux connecting it to one electron. The non-crossing law is in effect always. Iron has rings of protons only because at the time of fusion creation, the electrons force the protons into lines of protons. This is a new concept to science, merging electromagnetism, the strong nuclear force concepts, Lenz's

Law, the Pauli Exclusion Principal, and material science into a harmonized reality. All these phenomena are at play deterministically forcing the elements to have the properties that we enjoy on our cool planet's surface. At no time are electrons independent. Even when the dark energy situation has a halo of electrons speeding away at the fringes of the universe, they left behind an equal number of protons as dark matter. Even those pairs are connected across billions of light years. Dark matter is matter with no local electrons. Dark energy is electrons at the far ends of the universe, sent there earlier during eternity, without protons.

Electrons in the supernovas are deterministically paired with protons in the supernova at all times by lines of flux. Three dimensions connect one proton with one electron. If an ohmic cycle occurs due to two lines passing through each other, a new pair loyalty is created instantly, so at no time was any one electron independent of a proton. Pair loyalties can be changed, so a different proton has the line of flux to a different electron. In a superconductor at low temperatures, one electron can travel in a circuit for meters, while its line of flux is never broken. It never crosses another line as it makes loop after loop of the superconductor ribbon. The flux is a membrane of zero thickness. It has two space-like dimensions that can stack up without limit, taking up no thickness. But if the temperature gets too warm, the non-crossing law allows an ohmic cycle to occur, dissipating heat when a new loyalty is established to a different local proton, far from the original proton that was paired with that one electron. This implies that in a star at a billion degrees, iron's rings are formed even though loyalties are changed. Flux lines do cross at high temperature, but it takes no time to do that. The electron at any instant is forcing the proton to be excluded from deviating its position out of the line. Twelve protons form a ring and twelve electrons are stacking up their flat membranes with no thickness so protons do not repel protons in the nucleus. They are forced by only two dimensions, not three. Coulomb repulsion is not like gravity with a three-dimensional force, when so close to a nucleus. Flat membranes get stacked without repelling each other in that one dimension. So, the protons group into a line with one dimension, enforced by the electrons. Future science is now revealed, today, June 7, 2021. Be certain that this is true. Irons has two rings of protons that are coaxial. Each electron is paired using a line to one proton. The lines do not cross each other unless a new loyalty is established instantly. These are laws of physics.

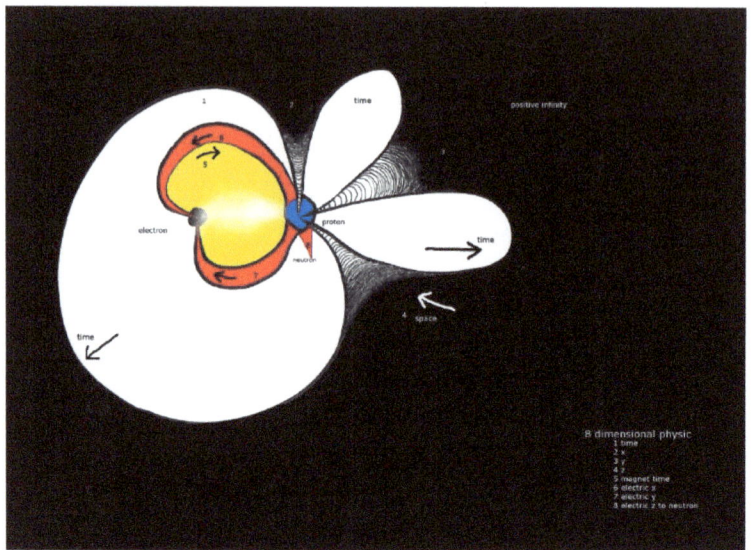

Figure 62: Eight dimensions and matter

Figure 39 shows one proton paired with one electron by two space dimensions hx and hy while the third space-like dimension hz goes to a neutron. Four gravity dimensions x, y, z and t are also shown. Three space dimensions go into the proton for gravity so **time grows** out [19C]. The electron has two space-like dimensions going in while one time-like dimension th, grows outward to the negative infinity in the proton. A mirror in the proton reflects the eight dimensions. A mirror in the electron reflects three dimensions: hx, hy, and th (magnetime). In your city on Earth, you have a silvered mirror. How deep is a mirror? A mirror is as deep as an electron is wide. A molecular wavefunction has three dimensions hx, hy, th. A photon has two dimensions hx and th. A photon is a piece of a wavefunction. Radio is not photons. Radio is induction. Radio is a cascade of motions of wavefunctions even in the partial vacuum between the stars. The matter field between each electron proton pair conducts the radio signal. Photons are flat and they slip through some molecules, being diverted, but not always absorbed. If a photon is absorbed into a molecule or ion, those two dimensions will cause a replenishment to occur for the third dimension of the wavefunction. In standard physics, electrons are usually responsible for photon emission and absorption. In this new theory both protons and electrons as a pair are responsible.

In Figure 39, the black represents potential space, where time has not reached yet from the blue proton. The one blue proton in this picture is emitting time as a white area. Three space dimensions, x, y, z, are shown as funnels being consumed by the proton. The space is like a liquid, moving into the proton. When a proton consumes its own volume of space from its surrounding, the proton has emitted 5ns of time. A neutron shown next to the proton as a label. A red dimension hz goes to the neutron from the proton. Two red dimensions, hx and hy, go from the proton to the electron. One white zone goes from the electron back to the proton and it is magnetime, th.

Four dimensions are for electromagnetism (Fig. 62). Three red dimensions (hx, hy, hz) go from the proton to the electron. One white dimension, th, goes from the electron to the proton. Four dimensions are for gravity. Three black dimensions, (x, y, z) are shown as tornado shapes. Time is emitted from the proton and it grows toward infinity, moving at the speed c toward the empty, black potential space. A mirror in the proton reflects the three black dimensions and the three red dimensions emerge from the proton. Time grows out of the proton and magnetime falls in. It is the balance of nature.

Abundances of the elements

The Sun has many elements in it [7C]. One of the rarest elements is lithium. Iron is one hundred thousand times more abundant than lithium. In the new theory of nuclear structure, lithium is rare because iron is common. Iron is composed of a cube with six Li-5 nuclei fused onto the faces of the cube-3. The abundances of elements with an even Z are greater than the abundances of elements with odd Z atomic number. That is because the symmetry of the pyramidal cube structure causes that formation to be more likely to survive the initial motions.

Looking at the periodic table in this book (Fig. 23), one can see that the tips of pyramids on most elements can contain a lithium isotope. The three protons on the tips of most pyramids may be where lithium is consumed. If one fills in the chart of abundances so lithium would be as abundant as the average of carbon abundance and helium abundance, then lithium would have enough matter to supply the excessive abundances of silicon, iron, and cobalt, with some lithium left over to feed the construction of manganese nickel and copper. That is evidence that lithium's low abundance is due to it being consumed to make the heavier elements just listed.

Uranium-234 (Fig. 63) is an example for the nuclear structure that is rare but valuable for people. The internal positions of the protons have been a mystery to most scientists, until now. The positions are given, next. U-238 is shown in Fig. 65.

Figure 63: Uranium isotope 234 made with chocolate spheres

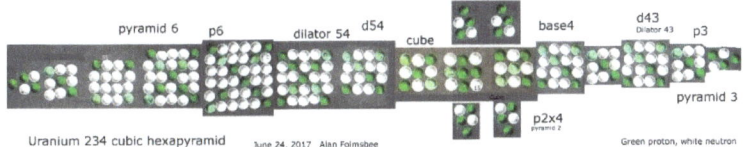

Figure 64: the internal allocations of protons and neutrons are now known in detail

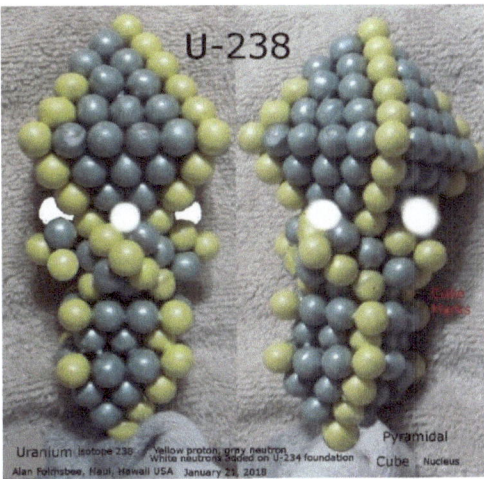

Figure 65: U-238 has four neutrons placed in the lowest pits on the surface between pyramids

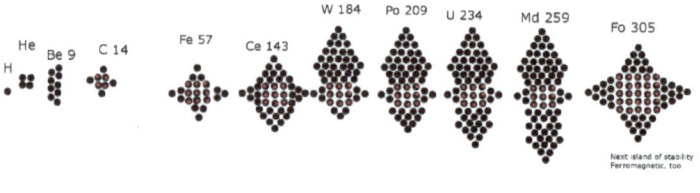

Figure 66: Uranium and other element profiles from June 23, 2017

In Fig. 64, one can count the 92 green protons and see where they are inside of uranium. The neutrons there are clear glass beads or marbles. Fig. 66 shows the date when uranium's structure was first predicted in a sketch of circular baryons, Jun 23, 2017. Baryons are protons and neutrons.

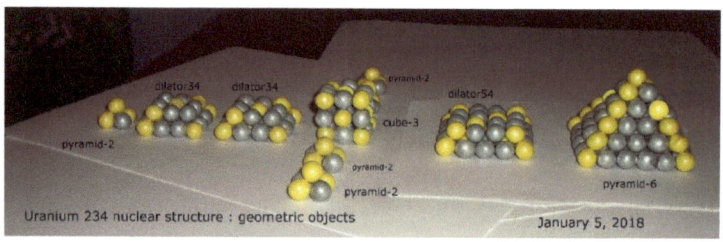

Figure 67: The modules for assembling uranium use integer geometries for this foundation element

Uranium is very rare in the Sun. It is even rarer than lithium by a factor of about 20. Helium is a billion times more abundant than lithium, which is notable because helium has two protons and lithium has three. Carbon is twenty million times more abundant than lithium, according to reference [7C]
https://www.amnh.org/exhibitions/journey-to-the-stars/educator-resources/our-star-the-sun/the-abundance-of-elements-in-the-sun .

That graph of abundances in [7C] can be compared with other sources of information, like reference [10C]
https://periodictable.com/Elements/003/data.html .
That shows lithium to the rarest element of all in the following link to a graph:
https://periodictable.com/Properties/A/SolarAbundance.html

There are many discussions about abundances of uranium and lithium in the scientific literature, but they were not aware of the static nucleus theory. Iron consumes lithium when it is fused in stars. That is why lithium is rare and iron is abundant. The pyramid-2 on the left is the Li-5 isotope (Fig. 67).

Origins of the cubic lattice

The simple cubic lattice at the core of all elements heavier than boron may be primordial. The stars today can also produce some cubic shapes by random coincidences. The triple alpha process has been a theory in the past for explain how carbon is formed. Please consider another way for carbon to form. A 2x2x2 cube of six neutrons and two protons can come from a time before expansion had proceeded very much. This would be the primordial source of cubes of all sizes.

This is the theory of eternal expansion. Eternity was already in progress before the alleged big bang, 14 billion years ago. Imagine 15 trillion years ago during eternity, when all matter was in a single lump, ten light years across. Expansion is always going on, but so slowly that it is easy to ignore. Expansion is due to protons that are not next to neutrons. A dimension called hz departs from each proton and does not feed a neutron. That hz is a space-like dimension that grows because it is a reflection of gravity that consumes space, like when a liquid is consumed and something is excreted.

As time went on, the spaces between protons and electrons grew. Those particles were in a simple cubic lattice in some regions and in a hexagonal close-pack in other regions of the universe. As expansion continued, cubic regions split apart into discrete cubes and rectilinear solids. No cubes were so small as 3x3x3 as we see in iron, today. Eventually, a cube 128x128x128 separated from larger blocks.

Later on, cubes with 4x4x4 baryons were separated. From those, the 3x3x3 and 2x2x2 cubes became discrete entities. Those are the primordial cubes which gave cores to iron and carbon. Heavier elements, like the foundation element nihonium also used the cube-3 by inheriting it from iron.

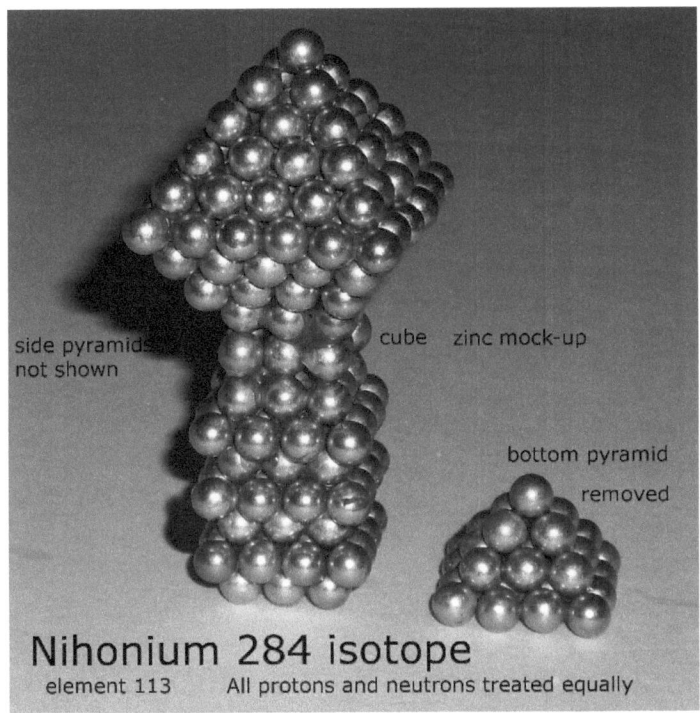

Figure 68: element 113 has a 3x3x3 cube at its core

The stacking of spheres (Fig. 45) is commonly using a nestling into the pits between the spheres. Heavier, radioactive elements have an elongated shape, making them vulnerable to fission and decay. This heavy element, Nh, is not naturally occurring so it is synthesized in small quantities in colliders. The question is: Where did the cube in Nh come from? A collision of plutonium and calcium can be use so element 94 and element 20 make element 114, which can decay to Nh, with 113 protons. Calcium is a cube-2 element and Pu has a cube-3. So, the answer is: the cube-3 in Nh came from an existing cube-3 in plutonium.

The question then becomes: Where does the cube come from in plutonium? Pu is not a naturally occurring element. It is manufactured in uranium reactors, where primordial cubes already exist or the cubes were made in stars. How do stars make uranium with the cube-3 at its core? Maybe neutron stars merge and make uranium. Maybe supernova cause fusion of something else to make uranium. Those are the common ideas that scientists have published in the past. But even those ideas can allow primordial cubes to be the ultimate source of uranium from iron's cube-3 structure.

Iron is one of the most abundant elements. Standard science claims lighter elements fused into iron. That may be, but the author is promoting an alternative theory. Iron may get its cubic core from larger cubes, not smaller elements. That was described at the beginning of this chapter in a process called eternal expansion which made small cubes out of a vast cubic lattice.

Eternity is still in progress. It always has matter. To summarize the events before the big bang, imagine 16 quadrillion years ago. The universe was a dense collection of protons, electrons and few neutrons. Inside the proton is a mirror at a place located at negative infinity. There is an equal number of protons and electrons. Expansion is slowly increasing the distance between particles. Some regions have a simple cubic lattice and some places have a hexagonal close-pack situation. Today's elements come from the cubic zones. The hexagonal zones cannot survive as elements and they became the hydrogen-1 of today. Cubic zones broke into smaller cube-4, cube-3 and cube-2 cores of today's elements. The cubes survived until now because each face of each cube was armored by a pyramid, like iron. Today's iron has inherited primordial cube-3 modules and six pyramids (Li-5) were obtained from stars to let the element survive. The lithium-5 is from the hexagonal zones of long ago.

At a time 900 trillion years ago, expansion was continuing. Matter's spacings expanded as time grew . Electrons separated from protons to make a plasma. A chain reaction made all ions in the universe emit a gamma ray simultaneously. That ray made electrons move far and fast. They went at almost the speed of time, c, to the far reaches of the small, expanding universe. They are called dark energy, today. Those lost electrons are still near the fringes of observability. The protons are in matter and dark matter elements, filling intergalactic space transparently. No photons can be absorbed by nuclei in dark matter because there are no local electrons near dark matter. The galaxies were made where electrons collided head on with electrons. That is where the dark matter became matter again for us and the stars. When the gamma ray storm happened, matter all became dark matter as electrons went away. The electron collisions long ago, near dark matter, slowed down the electrons and created matter of today's galaxies. Electrons that did not collide became dark energy at the perimeter of the universe.

Conclusion

The author asserts that this static nucleus theory can show the correct structures of the nuclei for all elements. Some elements were evaluated to a lesser degree than the elements that are shown in images in this book. The evidence is abundant that many properties of the elements are caused by nuclear structure. The proton positions are knowable for all elements by following the 19 known Rules that are printed in the chapter that follows the Introduction. Chemists can soon produce more accurate simulations and predictions of chemical reactions by using the exact proton positions and neutron positions in each isotope. The reasons for some nuclear decays can be seen in the shapes of some nuclei. A new age of science is now dawning.

This is a revolution in physics. There is resistance against this wide-ranging progress among some editors of journals. That is interpreted as an attempt to be efficient at rejecting publications that do not contain enough algebra. Geometry is the appropriate branch of mathematics to be used to describe the structures of the elements. Physics journal editors may want to prevent publication of simple solutions to areas of research that so many Ph. D. physicists have found intractable. It is hard to accept that an electrical engineer has succeeded where physicists have hit a dead end by using quarks. The entire area of quantum chromodynamics is declared to be obsolete. It is a wrong idea. Quark theories have never accomplished anything of practical value. The static nuclear theory will deliver vast wealth to people who use it in chemistry, nanotechnology, spintronics, superconductors, and to produce a new generator invention for making clean energy.

The dawn of the future of science includes unanswered questions about this frontier of physics. Why are the noble gases foundation elements? Where do the cubes come from? Are the cubes all primordial or are they created in stars today during random coincidences? Why are diamagnetic elements mainly in the eight groups on the right side of the periodic table? Why are light diamagnetic elements limited to H, He, Be, B, C, and N? Why does an element decay without a collision? Why does the commonest isotope of iron have one less neutron than the simplest shape (Fe-56 versus Fe-57)? Do proton currents have a category called a loyalty current where pairing with electrons changes?

There are some ideas to answer those questions about the frontier of physics, but the author cannot do this work alone. The heavy lifting is completed for the static nucleus theory. It is up to younger generations of researchers to produce more certainty among the students of the future. Join the revolution in science. Those who stand with the author will benefit. Those who are slow to accept the correctness of the integer geometries of nature will still be eligible to learn, later on.

The theoretical physics provided in this book is believed to be the correct theory of the shape of matter. The nucleus is the only permanent theoretical shape in the universe. These are the dreams that stuff is made from.

References Section C

1C Early periodic table published on October 11, 2017: **A Traction from Gravity to EM: Iron 56 (traction8d.blogspot.com) iron-56 page**

2C Neon simulations: The atomic nucleus: fissile liquid or molecule of life? (phys.org)
   https://phys.org/news/2012-07-atomic-nucleus-fissile-liquid-molecule.html#nRlv

3C Barium experiments Heavy barium nuclei prefer a pear shape (phys.org)
   https://phys.org/news/2016-06-heavy-barium-nuclei-pear.html

4C Radon experiments and Radium experiments StudiesOfNuclearPearShapesUsingAcceleratedRadioactiveBeams_Archive (liv.ac.uk)
   https://ns.ph.liv.ac.uk/~lg/papers/StudiesOfNuclearPearShapesUsingAcceleratedRadioactiveBeams_Archive.pdf

5C Uranium fission bimodal mass distribution Fragment mass distributions in the fission of heavy nuclei (ipen.br)
   See Figure 5 of https://www.ipen.br/biblioteca/2011/18038.pdf

6C Nuclear magnetic resonance for carbon-13 the background to C-13 NMR spectroscopy (chemguide.co.uk)
   https://www.chemguide.co.uk/analysis/nmr/backgroundc13.html

7C Abundances of the elements The Abundance of Elements in the Sun | AMNH
   https://www.amnh.org/exhibitions/journey-to-the-stars/educator-resources/our-star-the-sun/the-abundance-of-elements-in-the-sun

8C Iron sparks when hit Steel and Flint - Why Does Striking Them Together Produce Sparks? (quirkyscience.com)
   https://www.quirkyscience.com/sparks-from-steel-and-flint/

9C Spintronics Introduction to Spintronics (umd.edu)
   https://www.physics.umd.edu/rgroups/spin/intro.html

10C A and Z: mass numbers and atomic numbers: The Royal Society of Chemistry provides this https://www.rsc.org/periodic-table and another website gives them, also:
https://periodictable.com/Elements/003/data.html

11C The author's main website on nuclear structure, including the neutron powers essay
   Pyramids on Nuclei of Elements (pyramidalcube.blogspot.com)
   http://pyramidalcube.blogspot.com/

12C The author's website about electron pairing with protons, abstract units of measure
   Future Science – is going to get going on "The Generator" (wordpress.com)
   https://impuremath.wordpress.com/

13C The author's website on early fabrications of nuclear sculptures, with element 123
   ferronuclear
   http://ferronuclear.blogspot.com/

14C The author's website with the first essay on the static nucleus theory attempt
   A Traction from Gravity to EM: Iron 56 (traction8d.blogspot.com)
   http://traction8d.blogspot.com/p/iron-56_30.html

15C The author's website on the gravitational time constant of 5.1 nanoseconds
   Continuum Science: Planets and Stars Confirm 5.1ns tau (fcontinuumgravity.blogspot.com)
   http://fcontinuumgravity.blogspot.com/p/research-planning.html

16C The author's website on calculus for gravity volume theory, including units of measure
   Gravity Volume Theory (fcgravity.blogspot.com)
   http://fcgravity.blogspot.com/

17C The author's youtube channel
   Alan Folmsbee - YouTube

https://www.youtube.com/channel/UCKMzn4jBbjLFeUm4i0U354g

18C The author's 3D model files for 3D printers on the thingiverse.com website
Designs - Thingiverse
https://www.thingiverse.com/globemaker/designs

19C Passage of Time in a Planck Scale Rooted Local Inertial Structure
by Joy Christian who writes that time is growing
http://philsci-archive.pitt.edu/1932/1/time.pdf

20C Link to Periodic Table of the Elements, June 19, 2021 version from
Period_big_gk.jpg - Google Drive
https://drive.google.com/file/d/1C4viPNfHrqjiUqINtSXkevDhoo0-yVVh/view?usp=sharing

21C The Journal of Nuclear Physics, March, 2019 "Magnetism from iron's nuclear structure" by Alan Folmsbee

22C **Charge distributions on the nuclei,** ISBN 979 836349 5403
This is a 535-page reference book by A. Folmsbee showing the positions of baryon in all 118 chemical elements. Page 255 Rules. Self-published on the Amazon website.

Index

A mass number, proton plus neutron count   9, 16
Abundance of elements 41, 94
Actinide series in periodic table 41
Anti-ferromagnetism 54, 65
Argon-40 isotope 54
Atomic number Z, proton count 9, 13
Baryon is a proton or neutron 5, 7
Boron 54, 98
Ca-40 calcium isotope 63, 99
Carbon 7, 19
Charge distribution 21, 63
Co cobalt 16, 38
Cu copper 27, 38
Dart function of pear-shape barium 86
Dimensions are eight in number 92
Electron 10, 46, 57, 68, 89, 93
Ferromagnetic 6, 12, 28, 38, 60, 65
Foundation element 7, 9, 61
Gadolinium 6, 16, 38
Gold 21, 81
Half-life for Pm 74
Helium 94
Iron 5, 13, 18, 27, 97
K-40 potassium isotope 9, 12, 23, 28
Lanthanide series 41, 42
Line of flux 52, 65, 89, 91
Law of Lines of Protons, etc. 21, 67, 90, 91
Mass number A 9, 16
Membrane called flux 89
Mn manganese nonferrous 17, 38, 69
N nitrogen 9, 17, 33, 47
Ni nickel 27, 94
O oxygen foundation element 9, 47, 53
Pairing theory 89, 107
Periodic table of nuclear structure 44, 41
P phosphorus foundation element 9, 47
Pt platinum 6, 80, 83
Pm promethium proof of iron theory 9, 17, 24, 33, 42, 51, 74

Replenishment after photon absorption 92
Rules of nuclear structure 9
Stability 11, 25, 30, 36, 71, 74
Static nucleus theory 7, 9, 15, 27, 40
Tc technetium 20, 22, 33, 42, 56, 74, 79
Tungsten W foundation element 80
Tetrahexahedron 30, 38
U uranium 234 foundation element 9, 21, 39, 57, 94
V vanadium catalyst 6, 80, 83
Vacuum 19, 23, 85, 92
Xe xenon foundation element 9
Z atomic number, proton count 9, 13
Zr zirconium foundation element 9, 22

About the author

Alan C. Folmsbee, MSEE 1989 University of Connecticut at Storrs
- Master of science engineer who studied physical electronics, electromagnetics, and solid-state physics
- 15 patents assigned to Intel, AMD, and Sun Microsystems, including a secure microprocessor invention
- Speaker at International Solid State Circuits Conferences and one International Electron Devices Meeting in the 1980s on semiconductor memory integrated circuit designs
- Upon retirement, I promoted myself to become a scientist (2014)
- Discovered the shape of matter while living in Maui, Hawaii from 2014 until 2021
- Browse pyramidalcube.blogspot.com for the detailed periodic table of nuclear silhouettes
- Now residing in Connecticut from May, 2021 until September, 2023
- Contact by email: folmsbee@protonmail.com or phone 1-808-269-8893 during 2023 and 2024

www.ingramcontent.com/pod-product-compliance
Lightning Source LLC
Chambersburg PA
CBHW040316220526
45473CB00009B/2453